Absolute Beginners Illustrated Guide to Organic Chemistry

A Plain-English Illustrated Introduction to Molecules, Carbon, and the Chemistry of Everyday Life

Science for Curious Students/Adults

Cybersoft Publishing LLC

C. Louis-Charles, PhD

Organic Chemistry - Carbon: Why Organic Chemistry Runs the World

This book is intended for general educational purposes only. It is not medical advice and is not a substitute for consultation with a qualified healthcare provider. Always consult your physician for individual medical decisions.

First Edition

Published 2026

Contents

Introduction

Pick up the mug. Feel its warmth through your palms. That first sip of coffee, the shampoo you used this morning, or the ibuprofen you took-all are examples of organic chemistry in action. You've been living with it every hour, often without realizing it. That realization can make you feel curious and empowered to learn more.

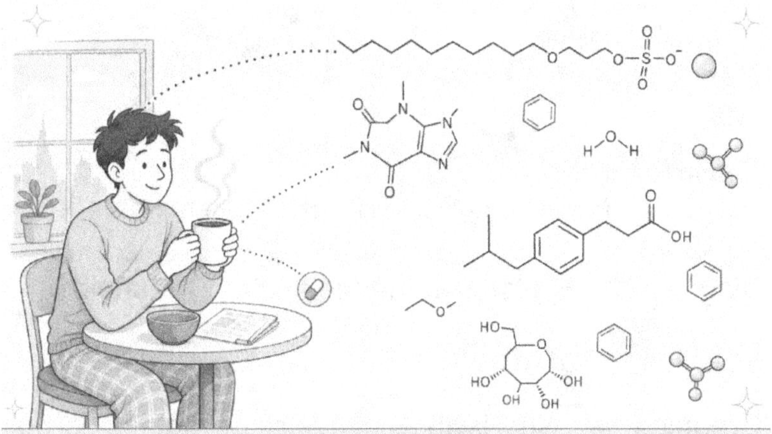

Every Morning Is a Chemistry Lab.

Here is what makes this particular branch of science so remarkable: it was never really locked away in a laboratory. It has been living in your kitchen, your medicine cabinet, your closet, and your bloodstream all along. The reason your coffee smells the way it does, the reason that ibuprofen finds the exact protein it needs and quiets your headache while leaving everything else alone, the reason your shampoo foams and your plastic water bottle never leaks, and the polyester in your jacket stays soft through a hundred washes. All of it is the same story, written in the same language, using the same alphabet. You just have not been given the key yet.

This book is the key.

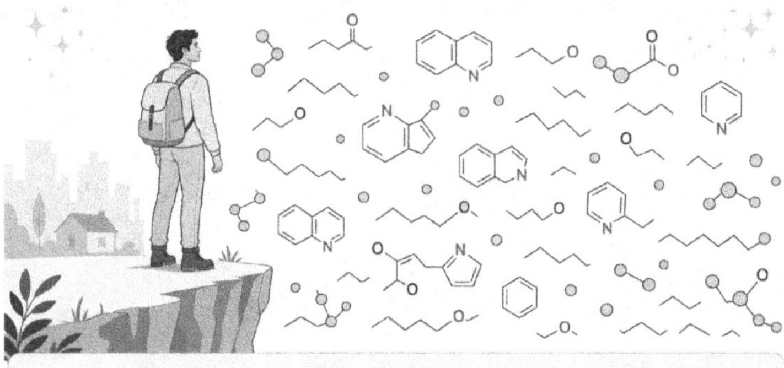

Explorer at the Molecular Threshold.

Diagram I.1: Explorer at the Molecular Threshold

What Organic Chemistry Actually Is

Organic chemistry is the chemistry of molecules built around carbon. That is the whole definition, and it is both simpler and more radical than it sounds. Carbon is not just one ingredient among many. It is the skeleton of almost every molecule that matters to living things: the fats in your cells, the sugars in your blood, the proteins that do the actual work of staying alive, the hormones that run your mood, the DNA that stores your entire biological history. Every drug in your medicine cabinet is an organic molecule. Every fuel that has powered human civilization is an organic molecule.

Every scent, every flavor, every dye, every plastic, every fragrance in a perfume bottle is an organic molecule. When chemists study organic chemistry, they are studying the operating system of the everyday world.

For most of the history of science, this subject was considered too complicated and too specialized for anyone outside a university laboratory. The textbooks made sure of that. They filled their pages with symbolic notation, reaction arrows, and mathematical machinery that was genuinely useful for specialists and genuinely impenetrable for everyone else. The result was a strange cultural gap: a science that explains almost everything a curious adult wants to understand, made inaccessible to almost every curious adult.

That gap does not serve anyone. And this book is not interested in maintaining it.

The Promise

By the time you reach the last page of this book, you will be able to read the molecular world. Not in a vague, hand-wavy, "I have a general sense of molecules" way, in a real way. You will look at the ingredient list on a shampoo bottle and recognize the functional groups that do the work. You will hear a news story about a new drug recall and understand, at the molecular level, exactly what went wrong. You will watch a cooking competition and know what the Maillard reaction is doing to that steak in the pan.

You are, at this moment, standing at the entrance to a hidden landscape. It is not hidden because it is exotic or distant. It is hidden because nobody handed you the map. That is what the next ten chapters are—a map. And the landscape it reveals is the one you are already living inside, made suddenly, astonishingly legible.

Think of yourself, from this page forward, as an explorer. Not a student grinding through material for an exam. Not a passive passenger being told facts. An explorer, moving through a real and consequential terrain, getting better at reading it with every mile.

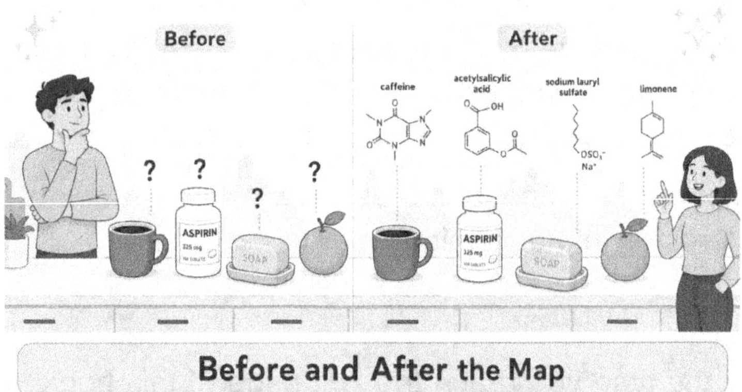

Diagram I.2: Before and After the Map

What Awaits: A Chapter-by-Chapter Preview

The journey opens with carbon itself, in "Carbon's Resume," where you will meet the one element that life chose as its backbone and find out exactly why. From there, "The Molecular Alphabet" hands you the actual visual language chemists use to draw and read molecules, turning an intimidating shorthand into something you will find obvious by the chapter's end. "Personality Chemistry" introduces functional groups, the molecular accessories that determine whether a compound smells like vinegar or nail polish remover, and whether it dissolves in water or repels it. Then "Mirror Molecules" takes you into the strange and consequential world of chirality, where two molecules can be identical in composition yet opposite in effect, and where getting it wrong once changed pharmaceutical regulation worldwide. "The Art of the Deal" tells the story of chemical reactions without a single symbolic arrow, using the language of attraction, collision, and rearrangement instead. "Designed to Heal" goes inside the drug development pipeline, from the serendipitous discovery of aspirin to the rational molecular engineering of modern medicine. "Miracle and Menace" follows the polymer, from the breathtaking invention of plastic to the molecular explanation of why a bag can outlast a civilization. "Taste, Smell, and the Chemistry of Pleasure" arrives just when you need a delicious chapter, decoding the molecular events behind coffee, perfume, and the Maillard reaction. "Burning and Beyond" follows the energy stored in carbon bonds from ancient sunlight to your gas tank to the biofuels and hydrogen cells of tomorrow. And "The Next Chapter" puts you at the frontier: green chemistry, biomimicry, CRISPR, and AI-accelerated molecular design, the places where organic chemistry is rebuilding the world it helped build.

Ten chapters. One molecular language. And by the end, a kind of fluency you did not think was available to you.

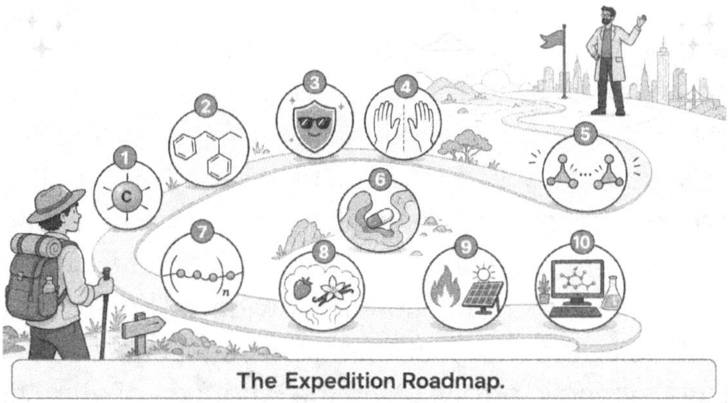

Diagram I.3: The Expedition Roadmap

A Word on Voice and Approach

You will not find math in this book. You will not find reaction arrows, IUPAC nomenclature strings rattled off without translation, or any of the symbolic machinery that makes organic chemistry textbooks feel like a foreign country. Those tools are genuinely useful for people who need to synthesize compounds in a laboratory. You do not need them to understand how organic chemistry works, and this book will prove that by never using them.

What you will find instead are analogies. Good ones, chosen because they do the actual work of making something invisible become visible. You will find more than 100 diagrams, all designed with consistent visual language that makes the molecular world navigable without prior training. You will find the direct address, because you are always in the room with this book, and there is no reason to pretend otherwise. You will find genuine respect for the difficulty of some of these ideas, and then immediate, practical effort to make them easier.

The voice in these pages is the voice of someone who finds organic chemistry genuinely extraordinary and cannot quite believe it is not already common knowledge. Not a professor reciting a lecture. A friend who read something astonishing and immediately called you.

Your intelligence is not in question here. Only your exposure. And that is about to change.

Welcome

The molecular world has been there all along, running quietly underneath everything you have ever touched, eaten, worn, smelled, or swallowed. It has grammar and logic and, once you can read it, a beauty that is genuinely hard to shake.

You are holding the map.

Let's go.

1 Carbon's Resume: Why One Atom Runs the Show

Look at your hand. Right now. Turn it over, flex your fingers, notice the way the skin creases at your knuckles. You are looking at the most sophisticated molecular architecture ever assembled in the known universe, and the overwhelming majority of it is built from a single element: carbon. Not a complicated element. Not a rare one. Carbon. The sixth element on the periodic table, number six, the one right between boron and nitrogen, the one that forms the tip of a pencil and the face of a diamond. That carbon.

You Are Mostly Carbon

The charcoal in your backyard grill is carbon. The diamond in a ring in a jewelry store window is made of carbon. The graphite in a mechanical pencil is carbon. The CO2 you are breathing out right now is carbon hitched to oxygen. Your muscle tissue, your DNA, the caffeine dissolving off your tongue from that cup of coffee you are holding, the coffee itself. All carbon. All the same elements, arranged in breathtakingly different ways.

This chapter is Carbon's job interview. By the end, you will understand not just what carbon is, but why it won. Why, out of 118 elements on the periodic table, did life choose this one? The answer is more elegant than you might expect, and considerably more surprising.

What is an Atom, actually?

Before we can talk about why carbon is special, we need to agree on what an atom even is. And I want to do this without invoking quantum mechanics, because quantum mechanics, as illuminating as it is, tends to cause eyes to glaze over before we even get to the interesting part.

Here is the version that will serve you with everything in this book. Imagine an atom as a tiny, nearly weightless cloud of space with a dense little kernel at the center. The kernel, called the nucleus, contains protons (which carry a positive charge) and neutrons (which carry no charge and are mostly there for structural stability, like ballast in a ship). Buzzing around that nucleus at very high speed, occupying the surrounding cloud of space, are electrons. Electrons carry a negative charge. In a neutral atom, the number of electrons exactly equals the number of protons, so the charges cancel out and the atom sits contentedly with no net electrical pull in either direction.

That is genuinely all you need—protons in the center. Electrons are swarming outside. The number of protons defines what element you are dealing with: one proton, hydrogen; two protons, helium; six protons, carbon. Change the proton count, and you change the element, full stop. Carbon is not a particular shape or size. Carbon is "six protons." That is its identity, its fingerprint, the thing that makes it irreversibly carbon and not something else.

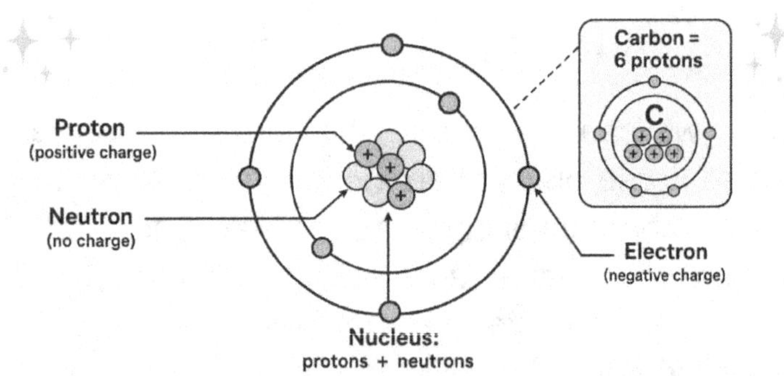

The Atom as a Solar System

Diagram 1.1: The Atom as a Solar System

Now, the electrons are the interesting part for chemists. They are the ones who actually interact with the world. They are the social layer of the atom. The nucleus sits quietly at home while the electrons go out and shake hands with other atoms, share resources, and form relationships. Those relationships are what we call chemical bonds, and chemical bonds are what build molecules, and molecules are what build everything you have ever touched, smelled, tasted, or been.

The electrons do not orbit randomly. They occupy layers, called shells, at increasing distances from the nucleus. The innermost shell can hold two electrons. The next shell out can hold eight. It is the outermost shell, the one exposed to the outside world, that determines how an atom behaves. Specifically, what matters is how many electrons the outermost shell is short of being full.

Atoms, it turns out, are desperate to have a full outer shell. It is the molecular version of wanting a complete set. If an atom is two electrons short of a full outer shell, it will cheerfully bond with two other atoms to borrow (or share) what it needs. If it has one electron to spare, it will bond once. If it has four to spare or four short, the math gets interesting, and that is exactly where carbon lives.

The Periodic Table as a Personality Chart

Most people remember the periodic table as a wall-sized chart in a high school classroom, full of boxes with strange abbreviations and numbers that seemed designed to resist memorization. That experience was unfortunate because the periodic table is not a memorization exercise. It is a personality chart—a map of how elements behave in a company.

The chart is arranged from left to right within each row in order of increasing proton count. But the rows themselves carry meaning. Elements in the same column, called a group, share a family resemblance.

They have the same number of electrons in their outermost shell, which means they bond to other atoms in roughly the same way and with roughly the same enthusiasm. Sodium and potassium, both in the first column (Group 1), both have one lonely electron in their outer shell, and both react explosively with water. That is not a coincidence. The chart is doing exactly what it was designed to do: organize elements by personality.

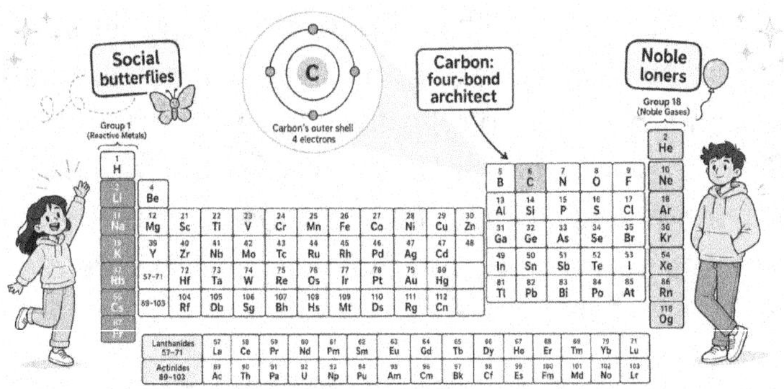

The Periodic Table as Personality Neighborhoods

Diagram 1.2: The Periodic Table as Personality Neighborhoods

Carbon sits in Group 14, the fourth column from the left if you are only counting the main-block elements. Its outer shell has four electrons. The outer shell can hold eight. So carbon has exactly four electrons and needs exactly four more to feel complete. That means carbon can form exactly four bonds.

Four. That number is going to come up over and over in this book, so let it sink in. Carbon bonds four times.

Not one, not two, not six. Four. And as we are about to see, four is a very particular kind of magic.

Why Four Bonds Change Everything

Think about a Lego brick. A standard eight-stud brick connects to other bricks on top, on the bottom, and on its sides. That versatility, the ability to connect in multiple directions and to multiple partners simultaneously, is what makes Lego useful for building complex, three-dimensional structures. If every Lego piece could only connect to one other piece, you could only make pairs. If it could connect to two others, you could make chains. With four connection points, you can build branching structures, rings, spirals, cages, and scaffolds. You can, in principle, build anything.

Carbon is the molecular version of a four-armed Lego connector. Each of those four bonds can attach to a different atom, or to another carbon. And because carbon bonds to carbon just as readily as it bonds to anything else, you can string carbons together in chains of any length, branch those chains in any direction, loop them back on themselves into rings, and keep building. There is no obvious upper limit. Vitamin B12, a molecule your body needs to make red blood cells and keep your nervous system running, contains 63 carbon atoms, all bonded together in an elaborate three-dimensional cage. The DNA in a single human cell, if you stretched it out, would be about two meters long. It is almost entirely a carbon backbone.

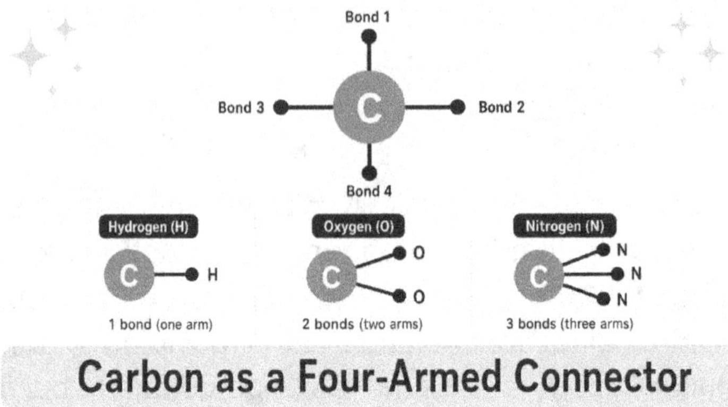

Diagram 1.3: Carbon as a Four-Armed Connector

Now compare that to oxygen. Oxygen is two electrons short of a full outer shell, so it bonds twice. Two bonds are fine for building simple molecules (water has two hydrogen atoms bonded to one oxygen), but they cannot form long chains. Bond to your left, bond to your right, and you are done. No branching. No rings. Nitrogen forms three bonds, which gives it more options than oxygen, but still lacks the architectural freedom that a fourth bond provides.

Hydrogen bonds once. Helium, neon, argon, and other noble gases do not bond at all. Their outer shells are already full, and they want nothing to do with anyone. They are the misanthropes of the periodic table, perfectly self-satisfied, chemically inert.

Four is special because it is the smallest number of bonds that allows for three-dimensional complexity. With four connection points arranged in space (they actually point toward the corners of a tetrahedron, a pyramid with four triangular faces), a carbon atom can build in every direction at once. It is not just a chain-maker. It is a scaffold-builder.

Here is a detail that will quietly blow your mind: the fact that carbon has exactly four bonds is, at a fundamental level, a consequence of quantum mechanics. The specific rules that govern how electrons arrange themselves in shells, rules that emerge from the strange but rigorous mathematics of wave behavior at tiny scales, land carbon precisely at four. This book is not a quantum mechanics book (that is a different title in this series, one that will take you somewhere equally surprising), but it is worth pausing to note that carbon's talent is not random. It falls out of the deep rules of the universe.

The Periodic Table's Other Four-Bonder

Now that you know four bonds are the key, you might wonder whether carbon is the only element that bonds four times. It is not. Silicon, directly below carbon in Group 14, also bonds four times. Silicon has the same number of outer-shell electrons. Silicon is far more abundant than carbon in the Earth's crust. Silicon is the backbone of computer chips. Silicon Valley is named after it.

So why is life carbon-based and not silicon-based?

It is a genuinely good question, and science fiction writers have been cheerfully exploiting the mystery for decades. The short answer comes down to three things.

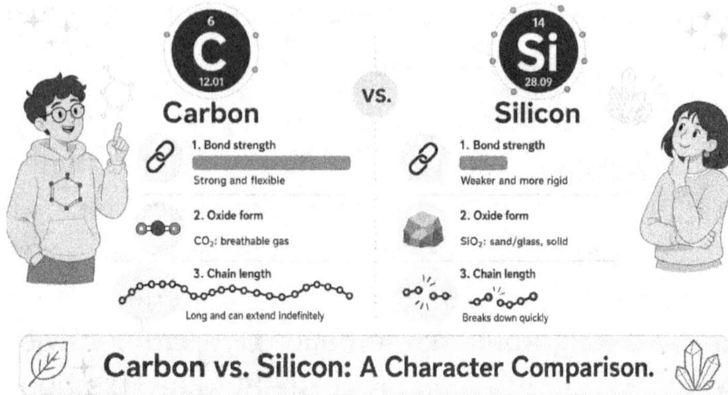

Diagram 1.4: Carbon vs. Silicon: A Character Comparison

First, bond strength and stability. Carbon-to-carbon bonds are strong, but also exactly the right kind of flexible. They can be single, double, or triple bonds, and each type has distinct properties that organic chemistry exploits constantly. Silicon-to-silicon bonds are weaker and have less double-bond character, which limits the variety of structures silicon can build.

Second, what happens when carbon or silicon burns? When carbon reacts with oxygen (combustion or respiration), it produces carbon dioxide, a gas. CO2 exits the body, exits the fire, exits the system. It is recyclable. Life can breathe it in and out. When silicon reacts with oxygen, it produces silicon dioxide, which is sand. Glass. Rock. Not a gas. Not recyclable in any biological timeframe. A silicon-based life form exhaling its metabolic waste would essentially be exhaling gravel, which is, at minimum, inconvenient.

Third, water. Life on Earth runs on water. Carbon compounds generally dissolve, react, and function beautifully in water. Many silicon compounds react with water in ways that destroy them. If you are trying to build chemistry in an aqueous environment, which every cell on Earth is, carbon is the far more cooperative partner.

Silicon is wonderful. It runs our digital world. But it did not get to build the biological one, and those three reasons above explain why.

Organic vs. Inorganic: The Line in the Sand

Once you understand that carbon bonds four times and can build essentially unlimited complexity, a natural question arises: what exactly counts as "organic chemistry," and what is left over for everyone else?

The historical definition of "organic" was, by modern standards, somewhat mystical. For most of the 18th century, chemists believed that the molecules found in living things could only be produced by living things. They called these molecules "organic" because they required a vital force, some spark of life, to create them. Inorganic chemistry was everything else: metals, minerals, acids, gases. Organic chemistry was the chemistry of life itself, and the two categories were thought to be forever separate.

That wall came down in 1828. A German chemist named Friedrich Wohler accidentally synthesized urea, a compound found in urine, from purely inorganic starting materials. He was not trying to make urea. He was experimenting with a completely different reaction and ended up with urea instead. He reportedly wrote to his mentor that he could "make urea without needing a kidney." The vital force hypothesis collapsed on the spot.

Diagram 1.5: The Fall of Vital Force: Wohler's Moment

Today, the working definition of organic chemistry is simply this: Organic chemistry is the chemistry of carbon-containing molecules. That is it. No mysticism. No vital force. If it contains carbon, it falls under the umbrella of organic chemistry. This includes everything from crude oil to aspirin to the plastic casing on your phone. Inorganic chemistry handles everything without carbon at its core: salt, rust, quartz, ammonia, and so on. There is overlap, edge cases, and compounds that sit on the boundary, but the carbon-or-no-carbon test handles 95 percent of cases without argument.

The term "organic" does not mean "natural" or "pesticide-free," a persistent source of confusion worth clearing up. A pesticide made of carbon compounds is organic in the chemical sense. A mineral supplement with no carbon is inorganic. The word is a chemical description, not a nutritional label.

Diamond and Graphite: Same Atom, Different Mood

Here is one of chemistry's most reliable crowd-pleasers. Diamond and graphite are both made entirely of carbon atoms. Nothing else. Just carbon. One is the hardest naturally occurring substance on Earth and the most coveted stone in jewelry. The other is so soft it smears onto paper and is used as a lubricant and a pencil filler. Same element. Completely different properties. The only difference is how the atoms are arranged.

In a diamond, each carbon atom is bonded to four neighbors in a tetrahedral pattern, forming a rigid, three-dimensional lattice that extends in all directions without interruption. Every atom is locked to four others, and those four are locked to four more, and so on, all the way through the crystal. There are no weak points. No layers that can slide. No easy pathway for anything to break through. This is why diamond scratches everything else, and nothing scratches it back.

In graphite, carbon atoms are arranged in flat hexagonal sheets, like chicken wire. Within each sheet, every carbon is bonded to three neighbors (not four), and the bonds within each sheet are actually stronger than the bonds in diamond. But the sheets themselves are held together by a much weaker force, a gentle attraction called a van der Waals interaction, which is roughly the molecular equivalent of two pieces of paper held together by static cling. When you write with a pencil, you are not digging graphite away. You are peeling sheets off and leaving them on the paper. The fourth bond electron in graphite is delocalized, spread out across the sheet, which also makes graphite a decent conductor of electricity. Diamond, where all four bonds are locked into place, is an insulator.

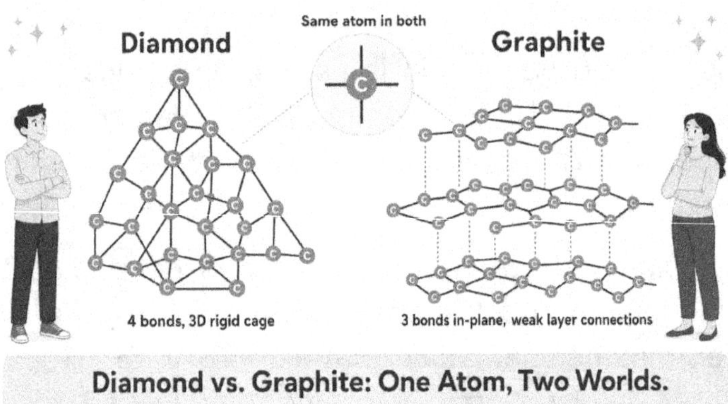

Diagram 1.6: Diamond vs. Graphite: One Atom, Two Worlds

The diamond-graphite comparison is not just a party trick. It is one of the clearest demonstrations of a principle that runs through all of organic chemistry: molecular architecture is everything. The identity of the atoms matters, but how they are arranged matters more. You are going to see that principle applied in increasingly sophisticated ways in every chapter that follows. For now, let it sit as a memorable proof of concept. Carbon can be a diamond or a pencil mark. The atoms decide nothing. The arrangement decides everything.

There is a third member of the carbon family worth mentioning: graphene. Graphene is a single sheet of graphite, one atom thick. It is two hundred times stronger than steel. It conducts heat and electricity better than copper. It is nearly transparent. It is also carbon—one layer of the same chicken-wire hexagonal arrangement, isolated from all the others. As of now, graphene is one of the most exciting materials in materials science, and it is just carbon with a new set of neighbors.

The Periodic Personality: Where Carbon Sits

To really understand why carbon became life's backbone, it helps to step back and look at where it sits on the periodic table with fresh eyes, not as a student trying to memorize atomic numbers, but as someone trying to understand a map of atomic personality.

The far left of the periodic table is where the most reactive metals live. Lithium, sodium, potassium. They have one lonely electron in their outer shell and are desperate to give it away. Drop sodium in water, and you get a small explosion. These atoms are the extroverts who need to connect so badly that they will bond to almost anything.

The far right is where the noble gases live: helium, neon, argon, krypton, xenon. Their outer shells are full.

They need nothing. They give nothing. In terms of forming bonds, they are entirely indifferent to the world around them. They pass through reactions without participating, like a tourist who speaks no local language and is perfectly comfortable with that.

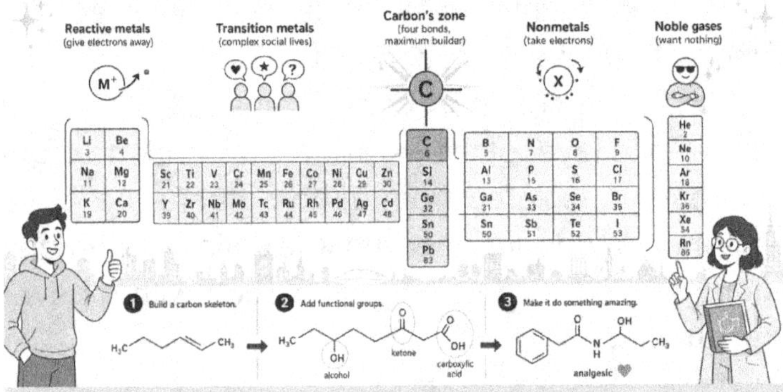

The Periodic Table Personality Spectrum.

Diagram 1.7: The Periodic Table Personality Spectrum

In between those extremes is where the chemistry of life happens. And carbon, sitting in Group 14, right in the middle of the periodic table's main block, is in the most architecturally useful position imaginable. It does not desperately want to give electrons away. It does not desperately need to take them. It wants to share, and it has four slots to share with. It is the collaborative, community-minded element in a periodic table full of more extreme personalities.

Carbon's neighbors in the same row also play important supporting roles. Nitrogen (one position to the right) forms three bonds and appears everywhere in proteins and DNA. Oxygen (two positions to the right) bonds twice and is the pivot of countless reactions. Hydrogen (sitting above everything, bonding once) is carbon's most common partner. These four elements, carbon, hydrogen, nitrogen, and oxygen, account for about 96 percent of your body's mass. Everything else is trace.

The Carbon Cycle: The Great Recycler

Here is something that might reframe how you think about breathing. Every time you exhale, you release carbon dioxide into the atmosphere. That carbon did not come from nowhere. It came from the food you ate. The food you ate was once a plant, or an animal that ate a plant. The plant built itself out of CO_2 it pulled from the atmosphere. Which means the carbon atoms currently leaving your lungs were, within the last year or decade or century, inside a plant, inside the soil, inside the atmosphere, possibly inside another animal.

Carbon cycles. It is the ultimate recycler. The word "cycle" here is not metaphorical. Carbon atoms circulate through the atmosphere, through plants, through animals, through soil, through the ocean, through geological processes, and back again, on timescales ranging from a few hours (your breath) to millions of years (fossil fuels forming underground from ancient organic matter).

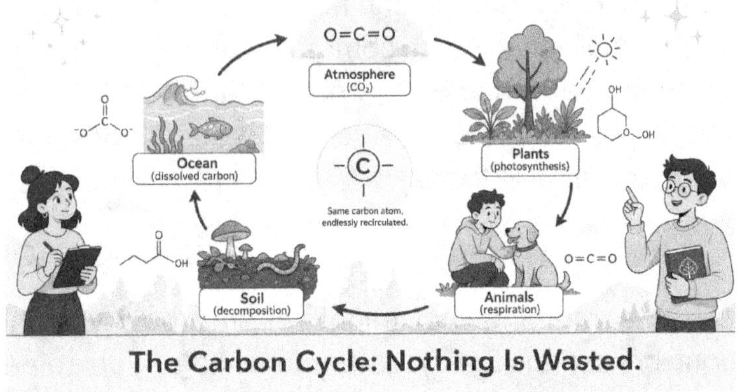

Diagram 1.8: The Carbon Cycle: Nothing Is Wasted

The carbon cycle matters beyond biology. The burning of fossil fuels, coal, oil, and natural gas, is essentially a case of humanity accelerating the geological part of the carbon cycle by millions of years. That carbon was locked safely underground as ancient organic matter. We are releasing it into the atmosphere in a geological instant. The carbon cycle, which evolved to handle carbon's slow movement through natural systems, is being stress-tested. Organic chemistry, in a very direct way, underlies the science of climate. You cannot fully understand either without understanding carbon.

For our purposes in this chapter, the carbon cycle is worth holding onto because it makes a counterintuitive point: the carbon in your body is borrowed. You are temporarily housing atoms that have been through many other forms before you and will continue cycling long after you. The carbon in your left hand might have been a leaf last spring. The carbon in your right might have been exhaled by a horse in 1850. There is something simultaneously humbling and wonderful about that.

Where Carbon Comes From: A Cosmic Origin

This is the part where the story gets genuinely awe-inspiring.

Carbon does not come from Earth. Not originally. It was not here when the planet formed. Earth, like all rocky planets, coalesced from a cloud of gas and dust about 4.5 billion years ago. That cloud was mostly hydrogen and helium, the simplest elements, forged in the first few minutes after the Big Bang. Carbon was not made in the Big Bang. It is too complex for that. Making carbon requires something much more extreme.

It requires a star.

Inside a star, hydrogen atoms are fused under conditions of pressure and temperature that are difficult even to imagine. This process, nuclear fusion, is what makes stars shine. As a star ages and its hydrogen supply runs low, the fusion process shifts. The star begins fusing helium atoms. And under precisely the right conditions, three helium nuclei can smash together and become a single carbon nucleus. This process, called the triple-alpha process, is what makes carbon possible.

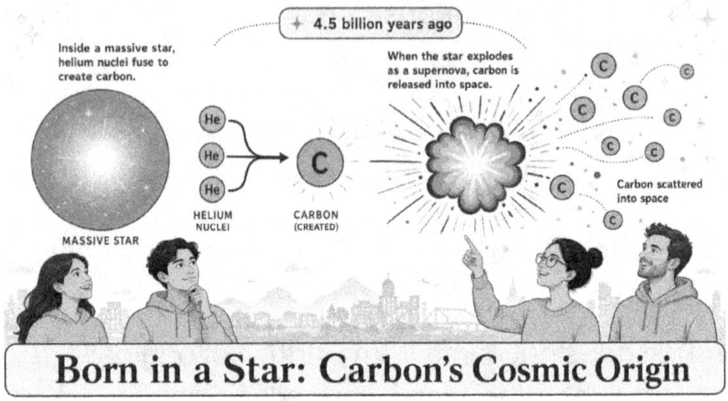

Diagram 1.9: Born in a Star: Carbon's Cosmic Origin

When massive stars exhaust their fuel, they explode as supernovae, scattering their contents, including all that carbon, across space in enormous clouds. Those clouds drift and gather under gravity. Some of them become new stars. Some of them become planets. Some of the scattered carbon ends up on a small rocky planet orbiting an unremarkable star in an outer arm of a spiral galaxy. Eventually, it ends up with you.

The carbon in your body is genuinely, literally, stellar. It was forged in a nuclear furnace at hundreds of millions of degrees, blasted across interstellar space, and assembled, over billions of years, into the shape of someone reading this sentence. If that does not produce at least a small, quiet shiver of wonder, I am not sure what would.

This connection between carbon chemistry and stellar physics is one of the things that makes science feel like one unified story rather than a collection of separate subjects. The same universe that produces stars also produces life, and carbon is the hinge between the two. If you want to go further into the cosmic story, the sibling title in this series on dark matter will take you even deeper into the astrophysics of what stars are made of and what fills the vast spaces between them.

Molecules: Carbon Meets the World

So far, we have talked about carbon mostly as a solitary atom with four open bonding slots. But of course, atoms never travel alone. They bond. They form molecules. And the simplest molecules that feature carbon give us a useful preview of what the rest of this book is about.

Methane, the primary component of natural gas, is carbon bonded to four hydrogens. Four bonds, four hydrogens, nothing fancy. It is the smallest possible organic molecule: a single carbon atom with all four bonding sites filled by the simplest possible partners. It burns, it smells (the "gas smell" is actually a different compound added to odorless methane as a safety warning), and it is the starting point for understanding how organic molecules are built up in scale and complexity.

Add one more carbon to methane, and you get ethane. Add another, and you get propane. Add a fourth, and you get butane. The lighter fluid in a disposable lighter is butane. The gas in a camping stove is propane. Keep adding carbons, and you move through the entire range of fossil fuels. Eight carbons, roughly, gives you the chain length found in gasoline. Sixteen carbons give you something that flows more slowly, like diesel fuel. Twenty carbons and above give you the thick, waxy chains in solid fats and candles. Same basic principle all the way down: carbon bonded to carbon, bonded to hydrogen, with the chain length determining the physical properties.

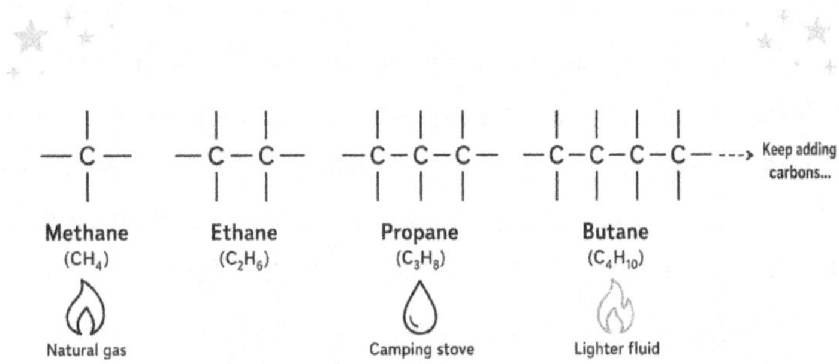

Building Up from Methane

Diagram 1.10: Building Up from Methane

The point here is not to memorize those molecules. It is to see the principle at work. Carbon chains are like a dimmer switch. Slide it one way, and you get a gas. Slide it further, and you get a liquid. Keep going, and you get a solid. The underlying chemistry is the same throughout. Only the chain length changes.

This is part of what makes carbon so powerful as a building block. You do not need different elements to get different properties. You need different arrangements. More length. More branching. A ring instead of a chain. A double bond instead of a single. Each architectural choice yields a different molecule with distinct behavior, smell, melting point, and effects on the human body. Organic chemistry is largely the science of understanding how those architectural choices produce those different outcomes.

Carbon in You: The Living Molecule

A human body is approximately 18 percent carbon by weight. That sounds modest until you remember that the largest component of your body (roughly 60 percent) is water, which contains no carbon at all. Strip out the water, and carbon suddenly represents nearly half of what remains. It is in every protein you build, every fat you store, every signaling molecule your cells use to talk to each other.

Your DNA, the instruction manual for every cell in your body, is a molecule built on a carbon backbone. The four letters of the genetic code (A, T, G, C) are not abstract symbols. There are four specific carbon-based molecules with specific shapes that fit together in specific ways. The information in your genome is stored in molecular shape, and molecular shape is a carbon story.

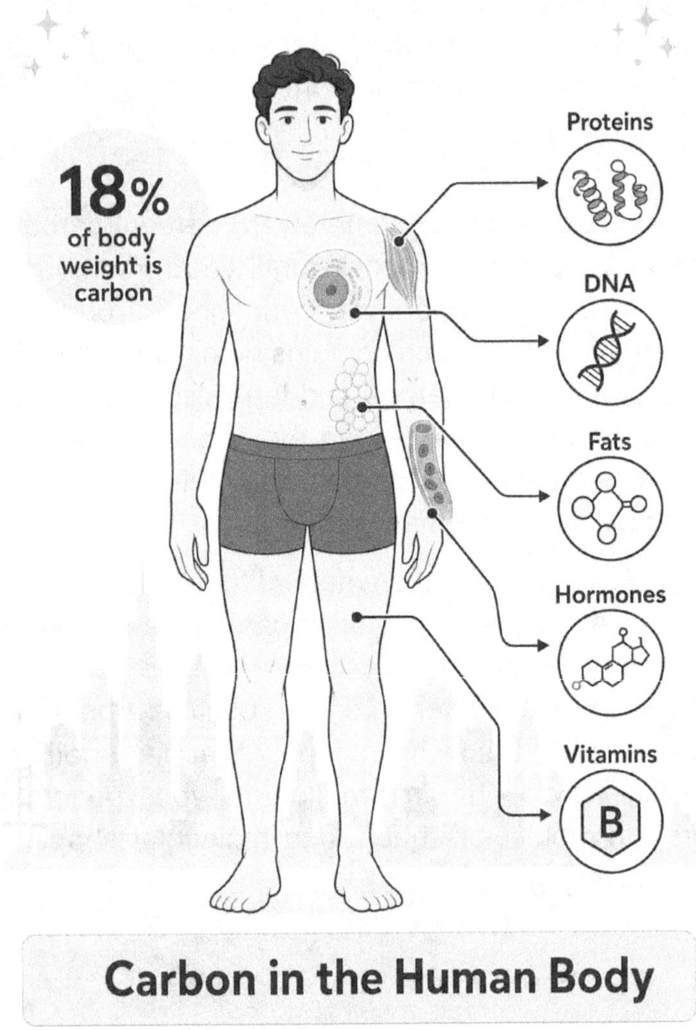

Diagram 1.11: Carbon in the Human Body

The hormones that regulate your mood, your metabolism, your reproductive cycle, and your stress response. All carbon molecules. The neurotransmitters that carry signals across the synapses in your brain right now, as you read and think and react. All carbon. The anti-inflammatory molecule your body produces after you bruise a knee. Melatonin signals your brain to go to sleep. The dopamine that gives you a small shot of satisfaction when something goes right. Each of these is a specific carbon-based molecular structure with a specific shape, and that specific shape is the reason it does what it does and nothing else.

Shape is the story. Shape is always the story. By the end of this book, you will be reading molecular shapes like a language, because that is exactly what they are.

Try This

Next time you hold a pencil, grip it for a moment and think: this graphite core is made of flat hexagonal sheets of pure carbon, stacking loosely, peeling off onto paper every time you write. The "lead" in a pencil has no lead in it. It never did (the misnomer dates to the 16th century, when graphite was first discovered in England and miners assumed it was a form of lead ore). It is carbon, in one of its most cooperative moods, leaving a trail of its atoms on every surface you drag it across.

Then look at any jewelry store. The diamonds in the window are the same element, arranged in a different mood: every atom locked to four neighbors in a cage that refuses to yield. Same atom. Different arrangement. Pencil mark and diamond, smear and sparkle. Carbon all the way down.

Takeaways

- You can explain what an atom is without using quantum mechanics: a nucleus of protons and neutrons surrounded by electrons, with the number of protons defining which element it is.
- You can describe why carbon forms four bonds and why that number, specifically, is what enables the construction of complex molecular architecture.
- You can explain, in plain terms, why life is carbon-based rather than silicon-based: carbon's oxide is a breathable gas, carbon chains are stable in water, and carbon has more structural flexibility at the bonding level.
- You can distinguish organic chemistry from inorganic chemistry: organic deals with carbon-containing molecules, inorganic deals with everything else, and the line between them is a practical guideline rather than a wall.
- You can describe how diamond and graphite can be made of identical atoms yet have opposite properties, because arrangement determines behavior, not identity.
- You can trace the origin of carbon in your body to stellar nucleosynthesis: the triple-alpha process inside aging stars, followed by supernova dispersal across space.
- You can explain the carbon cycle as the process by which carbon atoms circulate continuously through the

atmosphere, living things, soil, and oceans, making carbon simultaneously ancient and always in motion.

Into the Molecular Map

Diagram 1.12: Your Expedition Log: The Road Ahead

Now you know who the protagonist is. Carbon: six protons, four bonds, the element that builds chains, rings, and cages of unlimited complexity. The element that arrived on Earth aboard debris from exploded stars. The element that writes your DNA, fuels your cells, and leaves a soft gray mark on every page you have ever written on.

But knowing carbon exists is only the beginning. To do anything useful with that knowledge, you need to be able to read the molecular pictures that chemists draw to represent carbon's structures. Those drawings, called structural formulas or skeletal structures, have their own notation system, conventions, and visual logic. They look, at first glance, like a series of zigzag lines drawn by someone in a hurry. But they are actually a dense, efficient, beautiful language. Once you learn to read them, you will be able to look at any organic molecule and understand something real about it: how big it is, what it might smell like, whether it will dissolve in water, whether it might be a drug candidate, a pollutant, or a fuel.

That is where we are headed. Chapter 2 is your decoder ring.

2 The Molecular Alphabet

Pick up the aspirin bottle in your medicine cabinet, or a packet of sweetener, or vitamin C on your breakfast table. On the back of each is a chemical name, probably printed so small it takes a moment to find. You have always skipped past it. Here is why you are about to stop skipping.

You have seen them everywhere. On the chalkboard behind the barista, who takes her craft very seriously. On the opening credits of Breaking Bad, floating past the title like tiny hieroglyphics. In newspaper articles about new drugs, a spidery hexagon hangs in the corner of a graphic, and your eyes skip right past it. You have always known, on some level, that the hexagon means something. You did not know what.

Here is the good news: it is not a secret code. It is a shorthand, and like all good short hands, once you learn the rules, you wonder how you ever found it mysterious. A New York City subway map does not look like the actual streets of Manhattan. It does not show every building, every alley, every underground curve. It shows you what you need to navigate. The skeletal formula, that spidery thing full of zigzags and letters, is a molecular subway map. It shows you what you need. Nothing more, nothing less.

By the time you finish this chapter, you will be able to pick up a diagram of aspirin, caffeine, or vitamin C and read it the way you read a street sign. Not every detail, but the shape of the thing, the count of the atoms, the personality of the molecule. That is a skill most adults lack. You are about to.

Why are molecules hard to draw

Here is the first problem chemists faced when they started figuring out what molecules actually look like: molecules are three-dimensional objects, and paper is flat. A carbon atom does not sit at the center of a square. It sits at the center of a tetrahedron, a four-sided three-dimensional shape, with its four bonds pointing outward toward the four corners. Imagine holding a small ball in the middle of a room, and four strings run out from it toward four different spots on the ceiling, walls, and floor. That is carbon in space.

Drawing that on paper is genuinely tricky. You can try to show the 3D geometry by using special wedge-shaped bonds, where a solid wedge means "coming toward you out of the page" and a dashed wedge means "going behind the page away from you." Chemists do use those, and you will meet them in a later chapter. But for most purposes, showing every 3D detail makes the drawing crowded and hard to read quickly. If you are trying to communicate the structure of a molecule with twenty or thirty atoms, and you draw every single bond in its full three-dimensional glory, the picture becomes a tangle. The molecule stops looking like a molecule and starts looking like abstract art.

So chemists did what every smart profession does when faced with complexity: they developed a simplified notation that keeps the essential information and drops the rest. The result is the line-angle formula, also called the skeletal formula. And it is genuinely elegant once you understand what each part of it represents.

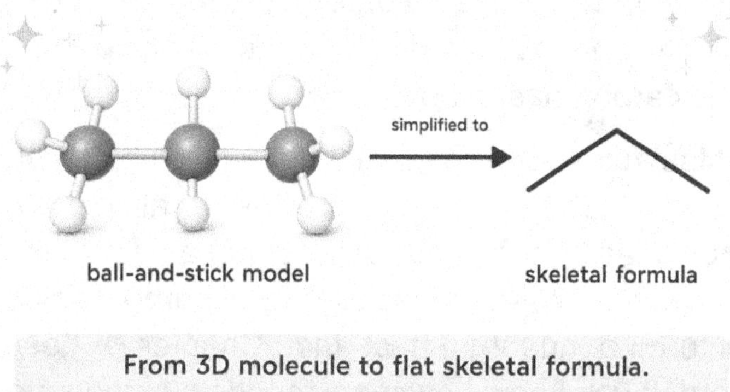

Diagram 2.1: From 3D molecule to flat skeletal formula

From 3D reality to 2D drawing

Think about what you actually need to know about a molecule to understand its behavior. You need to know which atoms are present. You need to know how they are connected, meaning which atom bonds to which. And you need to know if there is anything unusual about a particular bond, like whether it is a double bond rather than a single one. That is basically it. You do not need to know the exact three-dimensional coordinates of every atom in space, at least not for most conversations.

The skeletal formula captures all three of those things with astonishing economy. Every line segment represents one bond. Every angle, every endpoint where two lines meet, represents a carbon atom. Letters appear only when an atom is something other than carbon. Hydrogen atoms on carbon are not shown at all. They are assumed to be there, filling up whatever bonds carbon has left over.

That last rule is the one that trips people up at first. But there is a beautiful logic to it. Carbon almost always makes exactly four bonds. That is not a coincidence or a convention; it is a consequence of carbon's electron structure, a quantum fact that Chapter 1 gestured toward. If you know how many bonds a carbon atom is already making to other carbon atoms (or to the special atoms that do get labeled), you can always figure out how many hydrogens are hiding there. The formula becomes self-consistent. Once you internalize the rule, the missing hydrogens stop feeling like a mystery and start feeling obvious.

So when you see a zigzag line with six corners, you are looking at a chain of six carbon atoms, each with however many hydrogens are needed to bring their bond count up to four. No letters needed, no clutter, no confusion. The line says it all.

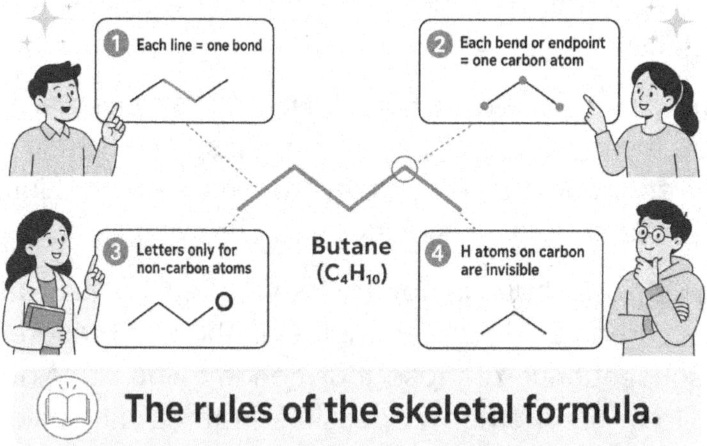

Diagram 2.2: The rules of the skeletal formula

Decoding the line-angle formula

Let us try it. Look at a simple case first: a three-carbon chain. In the skeletal formula, it is just a V-shape, two line segments meeting at an angle. The left end of the left line is a carbon. The angle where the lines meet is carbon. The right end of the right line is carbon. Three carbons total, no other atoms shown, which means the only other atoms present are hydrogens, filling out each carbon's four bonds.

The left carbon has one bond going to the center carbon. That leaves three bonds for hydrogens. Same for the right carbon. The central carbon has two bonds occupied, one on each side. That leaves two bonds for hydrogen. So the molecule has 3 plus 2 plus 3 equals eight hydrogens total. Propane. The stuff in your gas grill. A V with two lines, and you can read the whole molecule.

Now go up one step in complexity. Add another carbon. Now the chain is a zigzag with three segments and two angles. Four carbons, ten hydrogens. Butane. Lighter fluid. Same rules, a slightly longer drawing.

The zigzag shape is not arbitrary. In reality, the bonds in a carbon chain prefer angles of about 109 degrees, a consequence of those four bonds trying to point as far from each other as possible in three dimensions. When chemists flatten a carbon chain onto paper, the resulting 2D projection naturally looks like a zigzag. The drawing captures real geometry, translated into two dimensions.

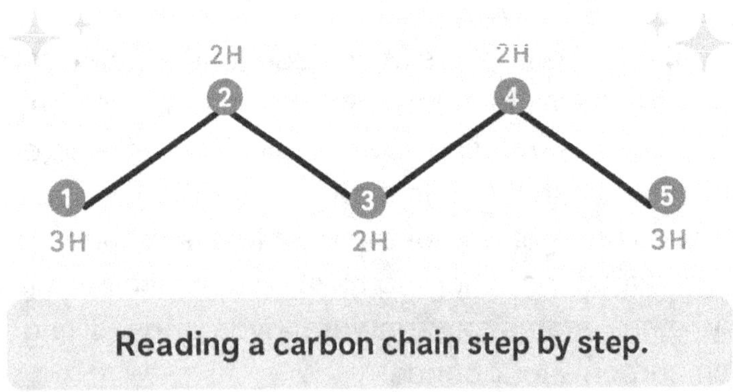

Diagram 2.3: Reading a carbon chain step by step

What happens when we introduce a double bond? In a skeletal formula, two lines drawn parallel and very close together between two atoms means those two atoms share a double bond rather than a single one. Highlighting this visual cue helps you recognize reactive sites, which are essential for understanding chemical behavior. The parallel pair of lines is striking and unmistakable once you know the convention. A double bond changes things: it uses up an extra bond connection on each participating atom, which means each of those atoms has one fewer spot for hydrogen. It also locks those two atoms into a specific geometry, but that story belongs to a later chapter.

For now, know that double bonds are drawn as double lines, acting as visual signals of reactivity. Recognizing this helps you feel confident in your understanding of chemical behavior, which is essential to your learning journey.

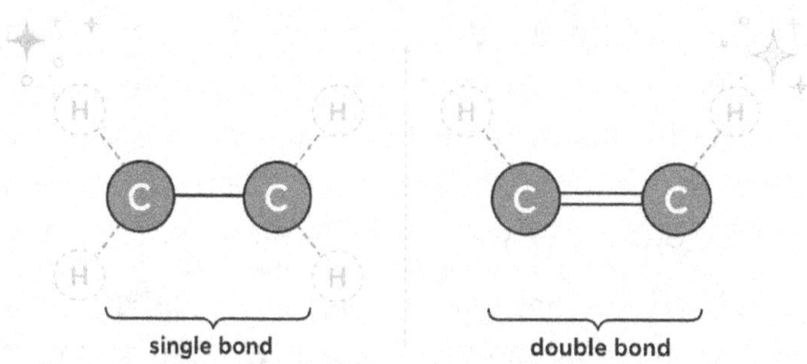

Single bond vs. double bond in skeletal form.

Diagram 2.4: Single bond vs. double bond in skeletal form

Aspirin: your first real molecule

Now, let us read something real. Aspirin is one of the most studied molecules on Earth, and its structural formula is a perfect first text in this new language. It has a hexagonal ring, a side chain with a double bond and an oxygen, and another arm with an oxygen and a carbon. Recognizing these groups, especially the benzene ring and ester group, is key to understanding its properties. It looks complicated at first glance. After this walkthrough, it will look like a sentence you can parse.

The hexagonal ring is benzene. That is such an important shape that it will soon get its own section. Still, for now, a regular hexagon in a skeletal formula represents six carbon atoms arranged in a ring, with alternating single and double bonds. Each corner of the hexagon is a carbon atom. The ring usually has three double bonds drawn inside it, shown as alternating single and double lines, or sometimes as a circle inside the hexagon. Either way, it is the same structure.

Aspirin has that benzene ring as its foundation. Coming off two of the ring carbons are two side chains. One side chain is an oxygen atom attached to a carbon atom, which is itself double-bonded to another oxygen atom and single-bonded to a CH3 group (that is, a carbon with three hydrogens, written explicitly because it is at the end of a chain). That ester group allows aspirin to release its active ingredient into your bloodstream. The other side chain is a carbon double-bonded to an oxygen (called a carbonyl) and then to an

OH group. That cluster is a carboxylate group, which makes aspirin slightly acidic and can irritate your stomach if you take it on an empty stomach.

You just decoded aspirin. You identified the ring, the two side chains, the key atom groupings, and their biological roles. That is organic chemistry literacy, and you did it in three paragraphs.

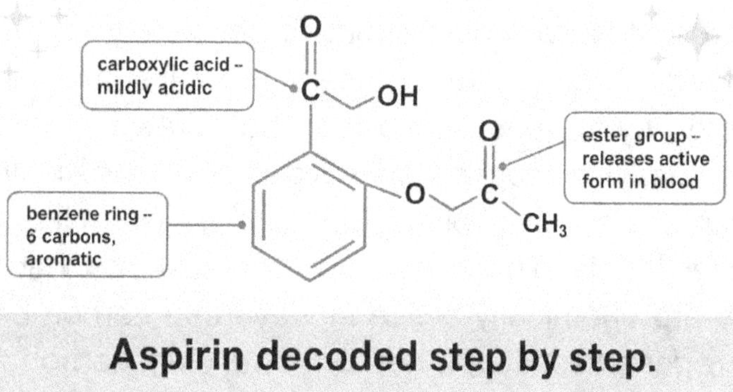

Diagram 2.5: Aspirin decoded step by step

Hydrogen: the invisible passenger

Let us linger on the absent hydrogen problem for a moment, because it really is what makes people stumble. When you look at a skeletal formula for the first time, and somebody tells you the hydrogens are invisible, the natural reaction is mild outrage. Why would you leave things out? Is that not cheating?

It is not cheating. It is editing. Think about the way a great novel works. The author does not describe every piece of furniture in every room. They give you the couch, the crack in the window, the smell of old coffee.

Your brain supplies the rest, and the story moves faster for it. In a skeletal formula, hydrogen is the furniture that you always know is there. Carbon always needs four bonds. If you count the bonds a carbon is making to its visible neighbors, the hydrogen count is simply four minus that number. This understanding helps you interpret structures quickly and accurately, trusting the notation to communicate efficiently.

Hydrogen is not unimportant. It absolutely matters in chemistry, and in later chapters you will see situations where hydrogens have to be drawn explicitly, particularly hydrogens on oxygen or nitrogen atoms, which appear in skeletal formulas as OH, NH2, and similar labels. Those hydrogens are shown because they are chemically active in ways that carbon-bound hydrogens usually are not. But on a plain carbon atom in the middle of a chain? The hydrogen is a given. It is the furniture. Your brain is doing the work, and that is the point of a good notation: it trusts you.

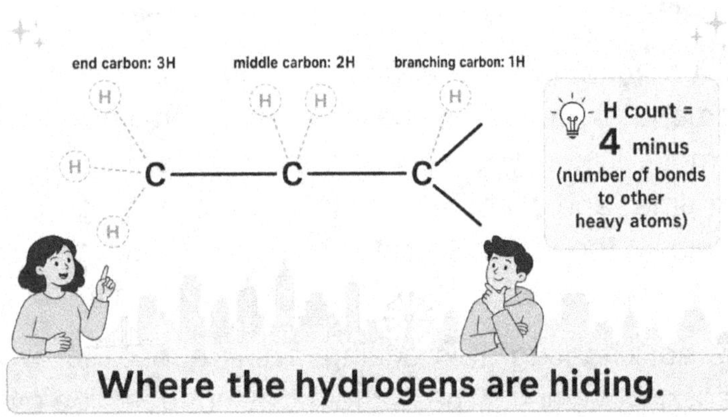

Diagram 2.6: Where the hydrogens are hiding

A gallery: familiar molecules in skeletal form

Here is where the new skill starts to feel genuinely useful. Recognizing molecules such as ethanol and caffeine in skeletal form helps you feel more confident when navigating familiar substances in your daily life.

Ethanol. The alcohol in wine, beer, and spirits. In skeletal form, it is almost laughably simple: a two-carbon zigzag ending in an OH group. Two carbons, the OH label on the end, and the whole thing is the size of a very short sentence. That OH group is the functional group that defines ethanol. It is why ethanol mixes with water (OH groups love water), why it enters your bloodstream the way it does, and why it metabolizes the way it does. Two carbons and an OH. That is the whole story.

Caffeine. Now things get a bit more interesting. Caffeine has a fused double-ring system: two rings sharing two atoms, resembling a pair of interlocking rings. One ring has five atoms, the other has six. Several nitrogen atoms appear, each labeled with an N. There are methyl groups (CH3, written explicitly) hanging off some of the nitrogen atoms. The overall structure is compact and dense-looking, which makes sense: caffeine is a biologically active molecule that has to fit precisely into certain receptors in your brain to do its job. We will talk about that locking mechanism in a later chapter. For now: spot the two rings, count the nitrogens (there are four), and notice that it is a denser, more complex structure than ethanol. The molecule's visual complexity hints at its chemical sophistication.

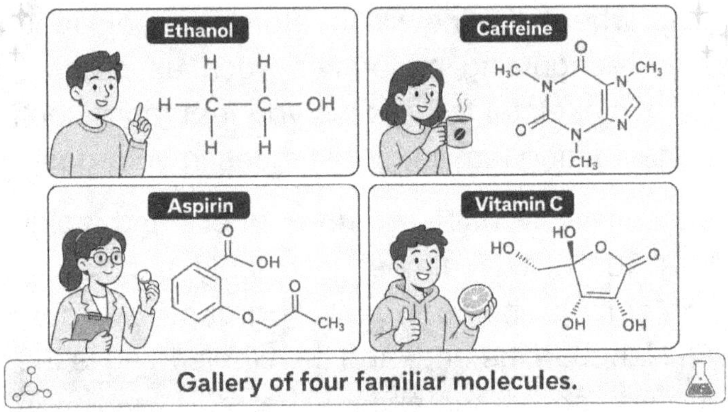

Diagram 2.7: Gallery of four familiar molecules

Vitamin C (ascorbic acid). A five-atom ring, mostly oxygen and carbon, with several OH groups hanging off it and an OH-terminated side chain. If you look at its skeletal formula, you will notice immediately that it is bristling with oxygen. All those OH groups are why vitamin C is highly water-soluble: oxygen loves water, and each OH group is essentially reaching out to grab water molecules and pull them close. You take vitamin C in aqueous tablet form, which dissolves instantly, and that is why. The structure explains the behavior.

Aspirin again, for comparison. Now that you have seen the others, aspirin looks like a middle-complexity molecule: a benzene ring at its core, which is more complex than ethanol's open chain but less complex than caffeine's fused rings. The two side chains with their oxygens are now obvious to you as functional groups. Recognizing these helps you feel more confident in reading molecules.

That is the point of this gallery. When you learn a new language, you do not start by reading dense philosophical texts. You start by recognizing simple words, then slightly harder ones, and one day you look up and notice you are reading a menu without translating it in your head. This is your first vocabulary list.

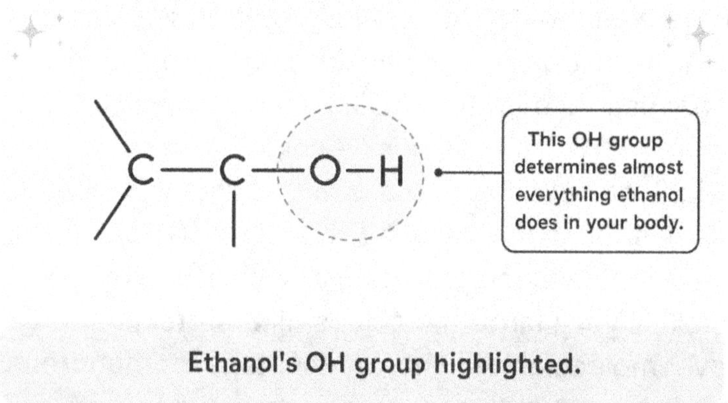

Diagram 2.8: Ethanol's OH group highlighted

Reading molecular weight and size

Once you can read a skeletal formula, you can start to extract more information from it. One of the most immediately useful things is a rough sense of molecular weight, which is just a number that tells you how massive the molecule is relative to a hydrogen atom. This helps you feel more in control of understanding molecular properties.

Molecular weight is not complicated to estimate. Each carbon atom contributes about 12 atomic mass units (AMUs), sometimes written as daltons. Each hydrogen contributes about 1 unit. Oxygen contributes about 16, nitrogen about 14. For molecules with multiple functional groups, count all atoms, including implied hydrogens, to get a reliable estimate. If unsure, break the molecule into parts and sum their contributions.

Why does molecular weight matter? Because it correlates, loosely but usefully, with things like boiling point, how easily a molecule evaporates, and how it behaves in the body. Small, light molecules tend to be gases or volatile liquids. Aspirin, with a molecular weight of 180 amu, is a solid at room temperature, heavy enough to stay put. Ethanol, at 46 amu, is a light liquid that evaporates readily, which is why you can smell it. Very large molecules, like proteins and DNA, have molecular weights in the tens or hundreds of thousands, and they are so massive that they do not evaporate at all.

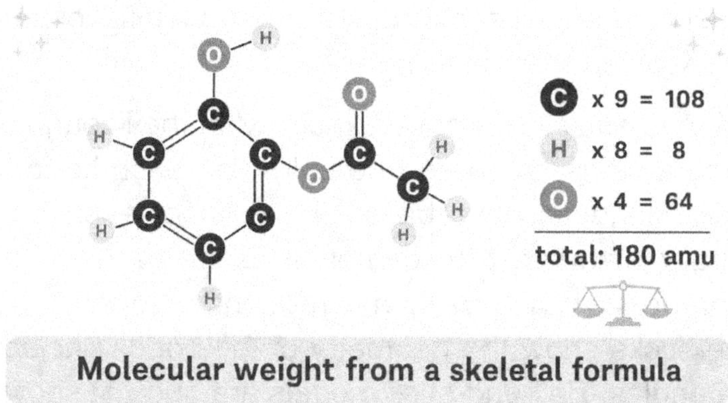

Diagram 2.9: Molecular weight from a skeletal formula

There is another dimension of molecular size worth understanding: how much physical space a molecule occupies. Two molecules can have similar molecular weights but very different shapes, and shape is enormously important in chemistry. Recognizing this can inspire curiosity about how molecules interact in real life, like how drugs fit into receptors.

For now, practice reading the size from a skeletal formula. Count the heavy atoms (everything except hydrogen). A molecule with six heavy atoms is small. Twenty to thirty heavy atoms is medium. Hundreds of heavy atoms are in a macromolecule. Thousands is a protein. You have just acquired a molecular scale.

Ball-and-stick vs. space-filling models

There is a second visual language for molecules that you will encounter whenever chemistry reaches into popular media, textbooks, or museum exhibits. It is the molecular model, and it comes in two main flavors: ball-and-stick and space-filling.

Ball-and-stick models are exactly what they sound like. Each atom is a sphere (the ball), and each bond is a stick connecting two balls. The color of each ball is usually standardized: carbon is black or dark gray, hydrogen is white, oxygen is red, and nitrogen is blue. The sticks show the geometry of the bonds, including the angles. Ball-and-stick models are good at showing connectivity and shape. They are not quite realistic in one sense: the sticks are not really there. Bonds are not physical rods. They are regions of shared electron density. But the ball-and-stick model gives you the molecule's architecture, which is enormously useful.

Space-filling models trade shape accuracy for volume accuracy. In a space-filling model, atoms are shown at sizes proportional to their actual atomic radii, and they overlap as in real molecules, crowding against each other with no visible sticks between them. The result looks like a cluster of spheres jammed together. Space-filling models show you how much room the molecule actually occupies and how it would appear to another molecule approaching it. They are used when shape and surface matter, for example, when discussing how a drug molecule fits into a receptor pocket.

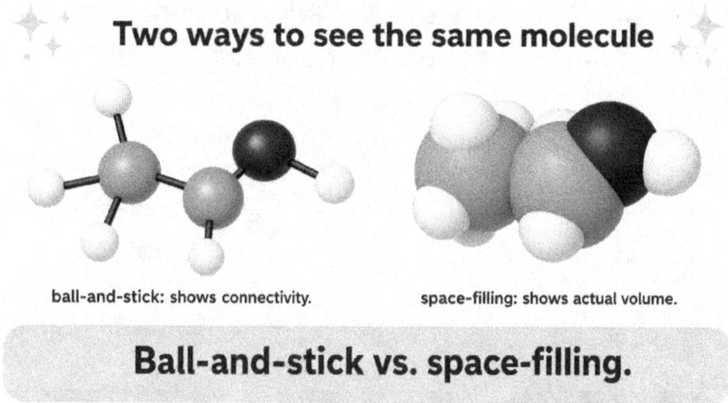

Diagram 2.10: Ball-and-stick vs. space-filling

Which model is most useful depends on the question you are trying to answer. A chemist asking "how does this molecule react?" often uses a skeletal formula, because it shows connectivity and functional groups most cleanly. To understand how a drug molecule fits into a receptor pocket, a space-filling model provides both volume and surface details. Combining models can give a comprehensive view of a molecule's behavior and interactions.

The skeletal formula is the most versatile of all because it is the quickest to draw and the easiest to modify. When chemists are working through a reaction mechanism or designing a new compound, they sketch skeletal formulas on whiteboards, in lab notebooks, in the margins of printouts. It is the working language of the field. Ball-and-stick and space-filling models tend to appear in publications and presentations when you want to communicate to a broader audience or show a specific feature of molecular geometry.

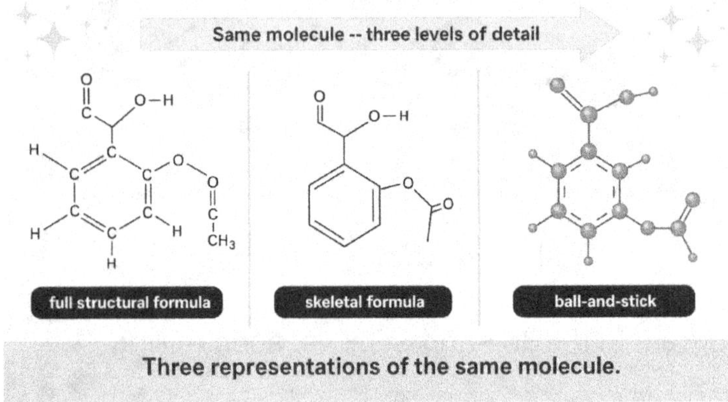

Diagram 2.11: Three representations of the same molecule

The subway map, revisited

Before we leave this chapter, let us return to where we started. The London Underground map is not a geographically accurate picture of London. If you projected it onto a real map of the city, the lines would curve and wander in ways the diagram does not show. Stations that look close on the Tube map are sometimes miles apart in reality. Everyone knows this, and no one cares, because the map does what it needs to do. It tells you which line to take to get from Paddington to King's Cross, and it does it without burying you in irrelevant detail.

The skeletal formula is a subway map of a molecule. It tells you which atom connects to which, where the reactive sites are, and what the overall shape of the carbon skeleton looks like. It does not tell you the exact bond lengths, the precise angles in three dimensions, or the electron density at every point in the molecule. Those details exist, and they matter in certain conversations, but they are not what you need for the conversation at hand.

Here is the thing about subway maps: once you have used one a few times, you stop thinking about how to read it. It just reads. The conventions disappear into habit, and all you see is information. That is where you are going with skeletal formulas. Right now, you are still conscious of the rules: endpoint is carbon, bend is carbon, letters are heteroatoms, hydrogens are implied. That is fine. That is learning a new alphabet.

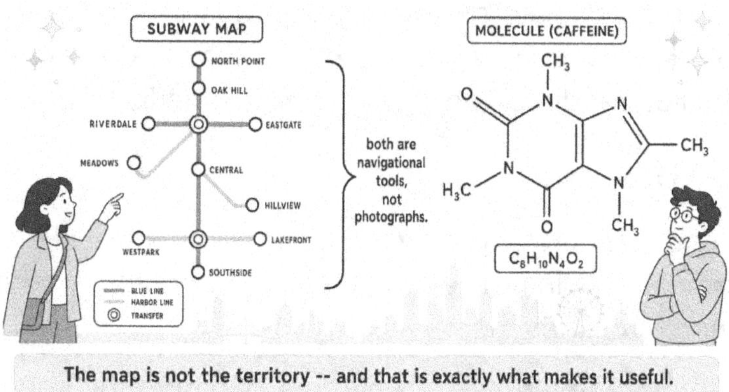

Diagram 2.12: Skeletal formula as subway map

After a few more pages of practice, those rules will drop below the level of conscious thought. You will look at aspirin and see aspirin, not a hexagon with side chains.

You will look at caffeine and see the fused rings, the nitrogen atoms, the methyl groups, and you will think: compact, nitrogen-rich, biologically active. The molecule will speak to you directly.

That is fluency. And you have just learned the alphabet.

Try this

The next time you pick up a bottle of medication, a supplement, or even a food package, search for the name of the active ingredient online and add the word "structure" to your search. You will almost certainly find a skeletal formula. See if you can count the carbon atoms (every vertex and endpoint), spot any atoms that are labeled (oxygen, nitrogen, and others), and make a rough estimate of how big the molecule is. You do not need to know what any of the groups do yet. Just practice reading. The meaning will come later.

Takeaways

• You can identify every carbon atom in a skeletal formula by finding every endpoint and every angle where two lines meet.

• You can determine the number of hidden hydrogens on any carbon by subtracting its number of drawn bonds from four.

• You can spot heteroatoms (oxygen, nitrogen, sulfur, and others) immediately because they are the atoms that get labeled with letters in a skeletal formula.

• You can read a double bond as two parallel lines between two atoms, knowing it means those atoms share more electron density and behave differently from single-bonded pairs.

• You can estimate the rough molecular weight of a simple molecule by counting carbon (12 each), hydrogen (1 each), and oxygen (16 each) atoms and adding them up.

• You can distinguish ball-and-stick models from space-filling models and know which type of question each one is best suited to answer.

• You can pick up the structural formula of a familiar molecule like aspirin, caffeine, or ethanol and navigate it without feeling lost.

Expedition log

You have the alphabet. You know how to read the lines and the letters, how to find the hidden hydrogens, how to sense the size of a molecule from its drawing. But an alphabet without vocabulary is just marks on a page. In Chapter 3, we meet the functional groups: the recurring chemical motifs that appear across thousands of molecules and give each one its personality. A carboxylic acid group makes a molecule acidic. A hydroxyl group makes it soluble in water. An amine group makes it basic and often biologically potent. Functional groups are the characters in every molecular story, and once you can recognize them on sight, the language of chemistry stops being a code and starts being a conversation.

3 Personality Chemistry: Functional Groups

Pour yourself a glass of wine. Go ahead. The molecule responsible for the pleasant warmth spreading through your chest is ethanol, a two-carbon compound with a particular arrangement of atoms tacked onto one end. Now imagine the same scene, except the bottle on your counter came from a hardware store rather than a winery. The liquid inside is methanol. It looks identical. It pours the same way. It even smells vaguely similar. But drink a tablespoon of it, and you will go blind. Drink a little more, and you will die.

Both are alcohols. Both carry the same chemical calling card, a tiny cluster of atoms that chemists write as -OH. And yet they are categorically different molecules with categorically different consequences for your body. The difference is not in that -OH tag. It is in the carbon skeleton underneath the body to which the tag is attached. Methanol's skeleton is one carbon. Ethanol is two. That single carbon means everything.

This is the chapter where you meet the wardrobe that molecules wear. The carbon skeleton, as we established before, is the body. The functional group is the personality: same body, different outfit, completely different character. And once you know the eight major outfits, you can walk into any room full of organic molecules, scan their silhouettes, and know how they are going to behave.

What a functional group is

Think of a mannequin. Stripped bare, every mannequin in a store looks the same: same shape, same proportions, same blank face. The mannequin is the carbon skeleton. Now dress it. Put a leather jacket on it, and it projects one set of signals. A lab coat and it projects another. A Hawaiian shirt, and suddenly the whole personality shifts. The jacket, the lab coat, the Hawaiian shirt: those are functional groups.

A functional group is a specific cluster of atoms, bonded together in a specific geometry, that is attached to a carbon skeleton. The keyword is "specific." The cluster is not random. It always has the same atoms, arranged the same way, and because of that, it always behaves the same way, regardless of what skeleton it is hanging off of. An -OH group on a two-carbon skeleton is ethanol. An -OH group on a six-carbon aromatic ring is phenol. Different molecules, wildly different applications, but they share family behaviors because they share that -OH group.

This is what makes functional groups powerful as a concept. You do not need to memorize a new set of rules for every molecule. You have learned the eight major groups and the behavioral logic of the vast majority of organic chemistry. Every drug, every flavor compound, every industrial solvent, every biological signal molecule can be understood, at least at first glance, by asking: which functional groups does it carry, and what does that tell me?

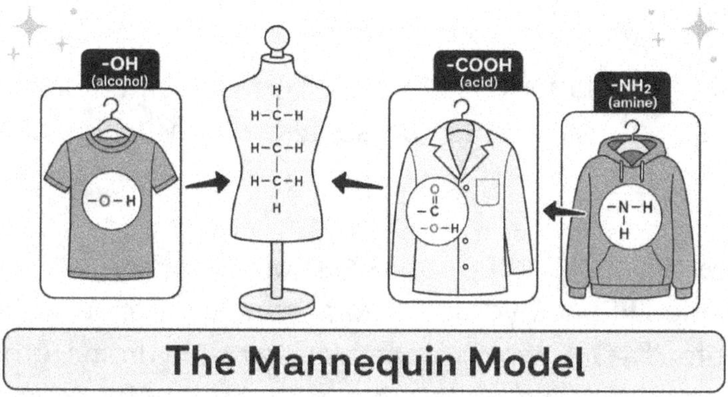

Diagram 3.1: The Mannequin Model

Chemists identified functional groups long before they fully understood why those groups behaved as they did. They noticed that molecules with similar "handles" acted alike: they smelled similar, reacted with the same reagents, and dissolved in the same solvents. The pattern was empirical first, theoretical second. Even today, when a medicinal chemist wants to change how a drug molecule behaves, the first conversation is almost always about the functional groups: which one to swap, which one to add, which one to strip away.

The number of functional groups is technically large. There are dozens of named ones in advanced organic chemistry. But here is the practical truth: eight groups cover the vast majority of what you will encounter in food, drugs, materials, and living systems. Eight personalities. Once you know them, you are fluent.

The eight personalities

Let us meet them one by one. This is not memorization for its own sake. Each personality comes with a story.

Alcohols (-OH)

The alcohol group consists of one oxygen atom bonded to one hydrogen atom, dangling from a carbon. Write it -OH, which chemists call a hydroxyl group. Oxygen is greedy for electrons, so the whole group has a strong partial negative charge on the oxygen side and a partial positive charge on the hydrogen side. That polarity makes alcohols social molecules: they mingle well with water, they hydrogen-bond with each other and with biological molecules, and they tend to be liquids at room temperature with higher boiling points than you would naively expect.

Ethanol is the alcohol you are most familiar with. Two carbons, one -OH. It dissolves in water easily (so your body can absorb it quickly), it crosses the blood-brain barrier, and it disrupts the signaling proteins that make your neurons fire. Methanol, one carbon and one -OH, is processed differently by your liver. The enzyme that metabolizes it produces formaldehyde as a byproduct, and formaldehyde destroys the cells of your optic nerve. Same functional group. Completely different skeleton. Catastrophically different outcome.

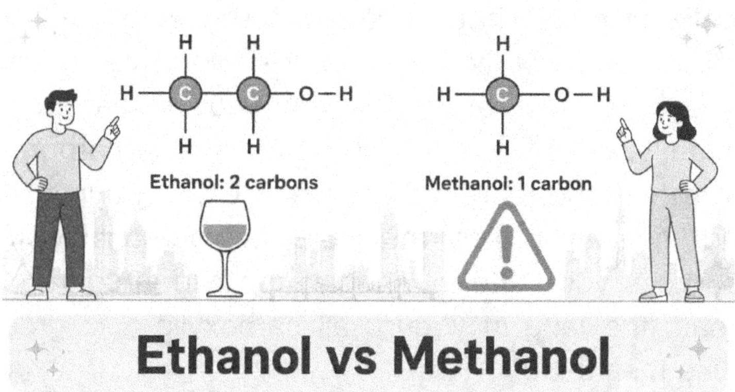

Diagram 3.2: Ethanol vs Methanol

Alcohols do not just appear in drinks. The rubbing alcohol in your medicine cabinet is isopropanol, a three-carbon compound with an -OH group in the middle. The glycerol in your skin lotion is a three-carbon chain with three -OH groups, which makes it extraordinarily good at absorbing water from the air and keeping your skin moist. Cholesterol, the molecule your doctor talks about in your annual checkup, is technically an alcohol, too. That one -OH group, buried in its complex ring structure, is a key part of how it docks into cell membranes.

Carboxylic acids (-COOH)

Now meet the sharp one. The carboxylic acid group is a carbon atom double-bonded to one oxygen and single-bonded to an -OH group. Write it -COOH, or draw it as -C(=O)OH. The double-bonded oxygen is called a carbonyl; the -OH sitting next to it makes the whole arrangement acidic, meaning it readily releases a hydrogen ion (a proton) into solution.

Acetic acid is what vinegar is. The smell hits you sharply and sourly because the molecule is small enough to evaporate easily, and your nose detects it instantly. Acetic acid is two carbons: one carbon, its three hydrogens, and then the -COOH group at the end. That is it. Tiny molecule, big personality. The sourness you taste in vinegar, the bite in sourdough bread, the tang in yogurt: all carboxylic acid groups doing their thing.

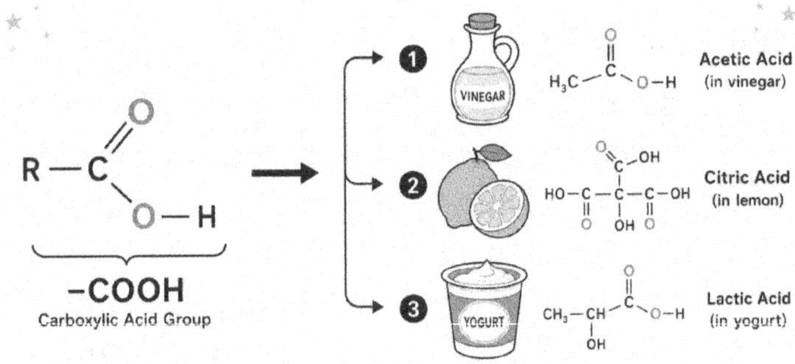

The Carboxylic Acid Personality.

Diagram 3.3: The Carboxylic Acid Personality

Carboxylic acids are biologically central. Amino acids, the building blocks of proteins, are named for their amino group (-NH2) and their carboxylic acid group (-COOH), both present in the same molecule. Fatty acids, the building blocks of fats and oils, are long hydrocarbon chains capped with a -COOH. The way your body stores and releases energy, the way it builds and dismantles proteins, all of that chemistry runs through the carboxylic acid group.

Amines (-NH2)

If carboxylic acids are sharp and sour, amines are the group that makes things smell like fish. An amine is a nitrogen atom bonded to one or two hydrogens, attached to a carbon chain. The simplest version is -NH2, one nitrogen with two hydrogens. Nitrogen, like oxygen, pulls electrons toward itself but less aggressively, making amines basic (they pick up protons rather than donating them) and often pungent.

The smell of rotting fish is largely a cocktail of amines with wonderful names like putrescine and cadaverine. These molecules are produced when bacteria break down proteins, and they are volatile enough to reach your nose from across the room. Your body uses many more refined amines for essential jobs: dopamine, serotonin, epinephrine, and histamine are all amines. The -NH2 group (or a close variant of it) is the piece of those molecules that interacts with receptors in your brain and body.

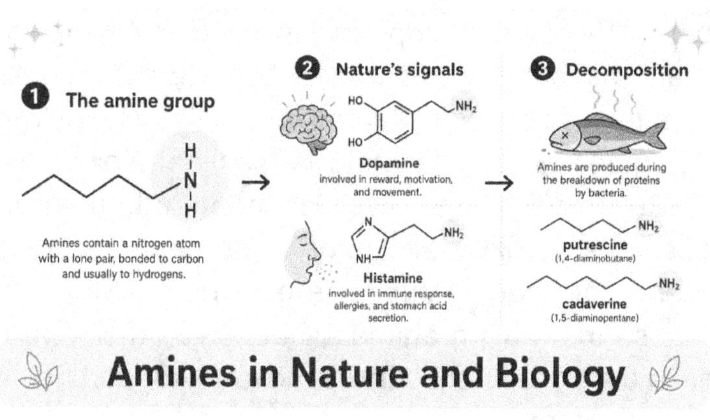

Diagram 3.4: Amines in Nature and Biology

Amines are also the backbone of drug design. The majority of drugs that act on the brain, antidepressants, stimulants, antihistamines, and local anesthetics, contain an amine group. That nitrogen is not decorative. It is the part of the molecule that threads into protein binding pockets, forms hydrogen bonds with receptor residues, and stays in solution long enough to be absorbed through your gut. Remove the amine group from many of these drugs, and you have removed the pharmacology.

Esters (-COO-)

Here comes the fruity one. An ester group forms when a carboxylic acid reacts with an alcohol, releasing water and linking the two through a -COO- bridge. The result is a group with one carbonyl carbon flanked by two oxygens. Esters are less polar than acids, which means they evaporate more easily, which means they tend to smell strongly.

Isoamyl acetate is the ester responsible for banana flavor. The same compound is used synthetically in banana-flavored candy and lip balm. Ethyl acetate is the ester in nail polish remover. Methyl salicylate is wintergreen. Benzyl acetate is jasmine. Nearly every fruit flavor you have ever encountered, natural or artificial, is a mixture of esters. The perfume industry depends on them. Your nose evolved to detect them because fruit-eating animals that could smell ripe fruit from a distance had a survival advantage, and ripe fruit is full of esters.

Diagram 3.5: The Ester Gallery

Esters are also how fats are stored. A triglyceride, the molecule your body uses to store energy in adipose tissue, is a glycerol backbone with three long fatty acid chains attached to it via ester bonds. When you digest fat, enzymes called lipases break the ester bonds, releasing the fatty acids for your cells to burn. The ester bond is also the linkage in polyesters, the fabric in your clothing. When a polymer chemist links acid molecules to alcohol molecules in a long repeating chain, the result is a plastic or a fiber held together entirely by ester groups.

Ketones (C=O in the middle)

A ketone is a carbonyl group, one carbon double-bonded to one oxygen, sitting in the middle of a carbon chain, flanked on both sides by carbons. That positioning matters. When the carbonyl is in the middle, you have a ketone. When it is at the end of the chain, you have something different (an aldehyde, coming next).

Acetone is the simplest ketone. Three carbons, carbonyl in the center. It is the main ingredient in most nail polish removers, a nearly universal industrial solvent, and one of the ketone bodies your liver produces when you are on a low-carbohydrate diet. If someone following a ketogenic diet has slightly sweet-smelling breath, that is acetone vapor. The carbonyl group in a ketone is polar. Still, the molecule lacks the -OH that would allow it to hydrogen-bond as strongly as an alcohol or acid, so ketones tend to be good solvents (they mix with water reasonably well) but also evaporate readily.

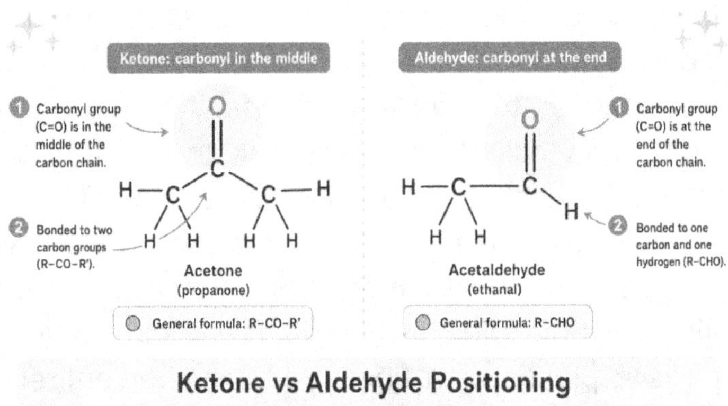

Diagram 3.6: Ketone vs Aldehyde Positioning

Cyclohexanone smells faintly of peppermint. Carvone is a ketone found in spearmint and caraway seeds. Many steroids, including testosterone and progesterone, have ketone groups built into their ring structures. The position of that carbonyl group is not just a label: it changes the electronic distribution of the entire molecule and governs how enzymes recognize and react with it.

Aldehydes (C=O at the end)

An aldehyde has the same carbonyl group as a ketone, but it sits at the very end of the carbon chain, with one hydrogen left on the carbonyl carbon. Write it as -CHO. That exposed hydrogen makes aldehydes more reactive than ketones, more prone to oxidation, and often more pungent.

Formaldehyde is the simplest aldehyde: one carbon, one oxygen, two hydrogens. It is a powerful disinfectant, a preservative used in embalming, and a known carcinogen at high exposures. It is also produced in small amounts by your own body as a byproduct of normal metabolism. Vanillin is an aldehyde: it is the molecule responsible for the smell of vanilla, and your nose detects it at extraordinarily low concentrations, which is why a drop of vanilla extract in a bowl of batter fills your kitchen. Cinnamaldehyde gives cinnamon its smell. Trans-2-nonenal, another aldehyde, is responsible for the "cardboard" flavor of stale beer and the particular smell of aging human skin.

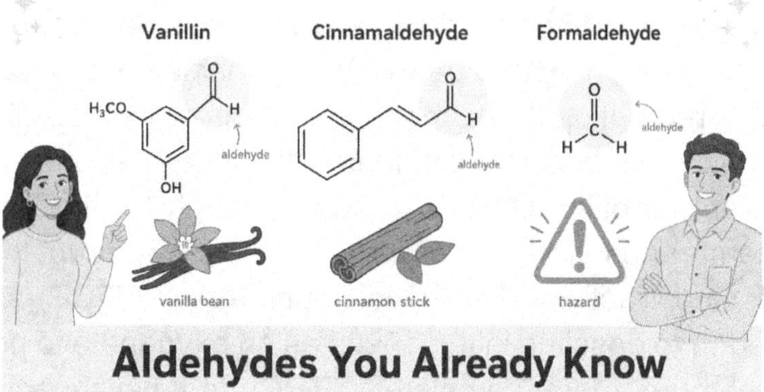

Diagram 3.7: Aldehydes You Already Know

Aldehydes are chemically interesting because they are at an oxidation midpoint. Alcohols can be oxidized to aldehydes, and aldehydes can be further oxidized to carboxylic acids. This is exactly what happens when wine sits open too long: ethanol gets oxidized first to acetaldehyde (that sharp metallic note in oxidized wine) and then to acetic acid (vinegar). The -OH becomes -CHO becomes -COOH, a three-stage journey all within the same carbon skeleton.

Ethers (-O-)

Ethers are the quiet ones. An ether group is simply an oxygen atom sandwiched between two carbon chains: C-O-C. No hydrogen on the oxygen. No double bond. Just an oxygen bridge. Ethers are relatively unreactive compared to everything we have met so far. They do not donate or readily accept protons and tend to be chemically inert under normal conditions.

Diethyl ether, with two ethyl groups flanking an oxygen, was the first widely used general anesthetic. Ether surgery parties in the 1840s were a genuine phenomenon, where people would inhale ether vapors for recreational numbness before its surgical application became standard. It works because its polarity profile allows it to pass through the lipid-rich membranes of nerve cells and temporarily disrupt their electrical activity. Methyl tert-butyl ether (MTBE) was added to gasoline in the 1990s as an oxygenate to help fuel burn more completely. THC, the psychoactive compound in cannabis, contains an ether linkage within its ring structure.

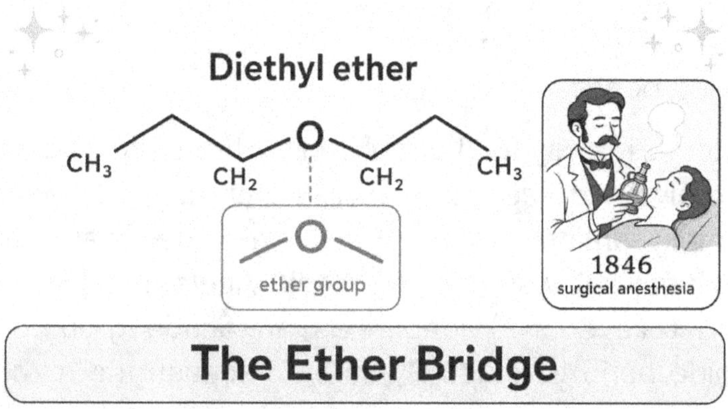

Diagram 3.8: The Ether Bridge

The polarity of ethers is intermediate. Oxygen adds some polarity compared to a plain hydrocarbon, but without an -OH group, ethers cannot hydrogen-bond the way alcohols can. They sit in an interesting middle zone: polar enough to dissolve some things that hydrocarbons cannot, nonpolar enough to dissolve things that water cannot. That dual-solubility character makes them useful lab solvents. Ethyl acetate, which we encountered as an ester, is often paired with diethyl ether in chemistry labs because together they can efficiently extract compounds from water-based mixtures.

Amides (-CONH-)

Save the most important one for last. An amide group is a carbonyl carbon bonded to a nitrogen: -C(=O)-NH-. It is, in a sense, the offspring of a carboxylic acid and an amine. When the two react and release water, the result is an amide bond. You already know this bond by a different name: the peptide bond. Every protein in your body is a chain of amino acids linked through amide bonds.

Nylon is a synthetic polymer whose backbone is entirely amide bonds. Penicillin and related antibiotics contain amide groups that mimic the shape of a bacterial cell-wall building block. Acetaminophen, the pain reliever in Tylenol, has an amide group. The amide bond is unusually stable because the nitrogen lone pair can delocalize into the adjacent carbonyl group, spreading electron density and making the bond resistant to hydrolysis under normal conditions. Enzymes called proteases have evolved specifically to break amide bonds, which is why digesting protein requires dedicated molecular machinery.

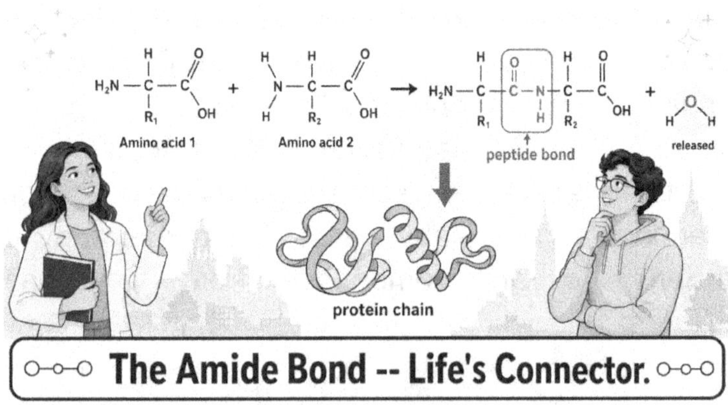

Diagram 3.9: The Amide Bond, Life's Connector

The stability of amide bonds is one reason proteins can maintain their structure under the relatively harsh conditions of a living cell. The amide bond is strong enough to persist while the protein does its work. Still, it is cleavable by specific enzymes when the cell needs to recycle or repurpose its molecular machinery.

Polarity, hydrogen bonding, and smell

Now that you have met all eight personalities, you have enough in hand to understand three of the most practically important phenomena in chemistry: why some things dissolve in water, why some things have high boiling points, and why things smell the way they do.

The short version is this: it all comes down to how electrons are arranged.

Polarity is the uneven distribution of electrons across a bond or across a molecule. When a carbon bonds to oxygen, the oxygen pulls the electrons toward itself, creating a partial negative charge on the oxygen side and a partial positive charge on the carbon side. The molecule has poles, like a tiny magnet. Polar molecules attract other polar molecules, and water is the most polar common molecule around. "Like dissolves like" is the chemist's shorthand: polar things dissolve in water, nonpolar things dissolve in fats and oils.

Among our functional groups, the ranking goes roughly like this. Carboxylic acids are the most polar and water-loving, followed by alcohols and amines, then esters and ketones and aldehydes (moderately polar), then ethers, and finally plain hydrocarbons (essentially nonpolar). That ranking tells you, without any calculation, whether a molecule will dissolve in a glass of water or float on top of it. A short-chain alcohol, such as ethanol, mixes completely with water. A long-chain ester, like the wax on an apple's skin, does not.

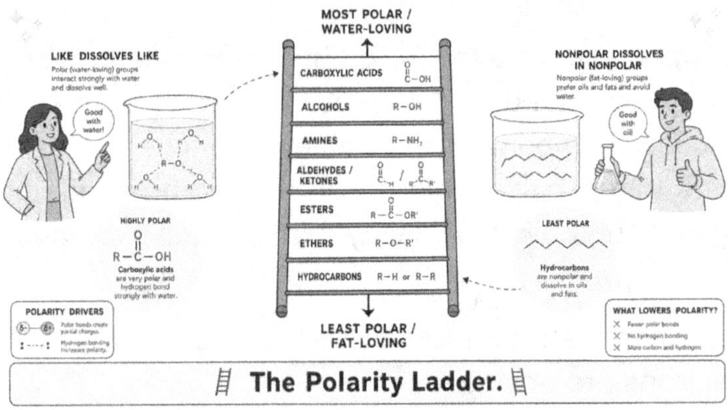

Diagram 3.10: The Polarity Ladder

Hydrogen bonding is a specific, stronger-than-average attraction between molecules. It requires two things: an atom with a lone pair of electrons (usually O or N) and a hydrogen bonded directly to an O or N (which makes the hydrogen slightly positive). When these two features are on different molecules, they attract each other, forming a hydrogen bond. It is not a full covalent bond; it is more like a strong handshake. But it is strong enough to meaningfully raise a molecule's boiling point and change its solubility.

Alcohols can hydrogen-bond extensively, which is why ethanol (two carbons) is a liquid at room temperature while propane (three carbons, similar size, no oxygen) is a gas. The -OH groups in ethanol molecules form hydrogen bonds with one another. Amines hydrogen-bond too, though less strongly than alcohols because nitrogen holds its electrons a bit less tightly than oxygen. Carboxylic acids can hydrogen-bond at both the -OH end and the carbonyl oxygen, making them particularly sticky molecules that often exist as dimers, two molecules hooked together, in solution.

Now for smell. Your nose detects volatile molecules, meaning molecules that evaporate and float through the air to reach your olfactory receptors. A molecule's volatility depends on its boiling point, which depends on how strongly its molecules stick to each other. Smaller, less polar molecules with fewer hydrogen bonds evaporate more easily and therefore smell more strongly. Esters are famously volatile and fruity. Aldehydes are volatile and sharp. Amines are volatile and fishy. Carboxylic acids are less volatile than you might expect given their polarity, but small ones like acetic acid (vinegar) still evaporate enough to reach your nose.

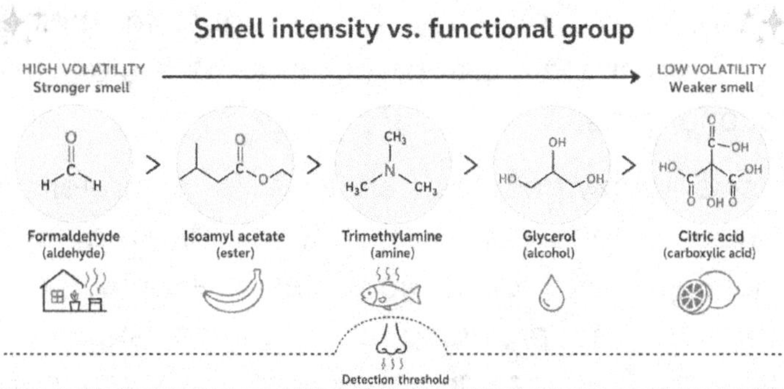

Why Things Smell -- Volatility and Functional Groups.

Diagram 3.11: Why Things Smell, Volatility, and Functional Groups

The size of the carbon skeleton matters too. As the chain gets longer, even polar molecules tend to lose water-solubility and gain fat-solubility because the nonpolar hydrocarbon portion starts to dominate. A short amine is water-soluble and fishy. A long amine is wax-like and nearly odorless. A fatty acid with sixteen carbons barely dissolves in water at all, while acetic acid with two carbons mixes with water in any proportion. Functional groups are personalities, but the skeleton underneath is the body they inhabit, and the body's size shapes everything too.

Functional groups as biology's active sites

The reason organic chemistry classes spend so much time on functional groups is not that chemists enjoy abstraction. It is that living systems evolved their entire molecular toolkit around functional groups. Every enzyme, every receptor, every signaling molecule, every cellular structure in your body is a story told in functional groups.

Consider enzymes. An enzyme is a protein that catalyzes a chemical reaction: it speeds up a reaction by holding the reacting molecules in exactly the right orientation and providing chemical assistance. The active site of an enzyme, the pocket where the reaction happens, is lined with specific amino acid residues whose side chains are functional groups. Serine proteases (enzymes that cut proteins) work by using a serine -OH group to attack the amide bond of the substrate. The -OH is the weapon. Without it, the enzyme is just a structural protein doing nothing interesting.

Drug design is almost entirely functional group management. When a medicinal chemist makes a new compound, they ask: Does it have an amine group that will protonate at physiological pH and stay soluble in blood? Does it have a carbonyl that will fit into the enzyme's binding pocket? Does it have a hydroxyl that will form a hydrogen bond with a specific residue in the receptor? These questions are answered by synthesizing molecules with precisely chosen functional groups and testing them against the biological target.

The concept of a pharmacophore is the minimum set of functional groups required for a molecule to have a particular biological effect. Strip away everything unnecessary, reduce a drug to its essential chemical features, and what remains is usually a handful of functional groups arranged in space. The functional groups are doing the biological work. The carbon skeleton is just the scaffold holding them in the right positions relative to each other.

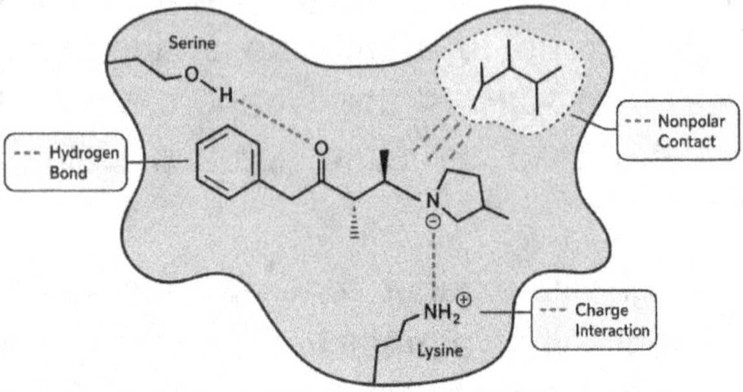

Functional Groups at an Enzyme Active Site

Diagram 3.12: Functional Groups at an Enzyme Active Site

The antibiotic penicillin provides a beautiful illustration of this. Penicillin kills bacteria by mimicking the shape and functional-group arrangement of a small bacterial cell-wall building block, thereby permanently blocking the enzyme that assembles the cell wall. The key functional group in penicillin is an amide bond inside a strained four-membered ring, a beta-lactam. That bond is so reactive, so willing to engage in chemistry, that it permanently attaches to a critical cysteine residue in the bacterial enzyme, turning it off permanently. The bacterium cannot build its cell wall, swells, and bursts. One functional group, precisely placed, stops a bacterial infection.

Same skeleton, different group, new molecule

Here is a thought experiment that ties everything together. Imagine a two-carbon skeleton: just C-C, two carbons, four total bonds available. Now dress it in different functional groups and watch the molecule's personality transform.

Put an -OH on one end: ethanol. Social, water-miscible, the molecule has been at the center of human social rituals for ten thousand years. Put a -CHO at one end: acetaldehyde, a sharp volatile compound produced when ethanol is partially metabolized in your liver, responsible for part of the hangover headache. Oxidize that aldehyde all the way: acetic acid, with a -COOH. Now the molecule tastes sour and smells of vinegar. Replace the -COOH with an -NH2: ethylamine, a simple volatile amine with a fishy, ammonia-like smell. Connect two ethanols through an oxygen bridge, and you have diethyl ether, the surgical anesthetic.

Same two-carbon backbone throughout. Different personality for each outfit. The skeleton did not change. The functional group did all the work.

This idea has profound practical consequences. Drug developers routinely practice what is called structure-activity relationship (SAR) research, in which they synthesize a series of molecules that share a carbon skeleton but carry different functional groups, then test each for biological activity. The goal is to find the precise combination of groups that gives the best therapeutic effect with the fewest side effects. Sometimes the difference between a useful drug and a toxic one is a single -OH group added to or removed from the scaffold.

The same principle operates in flavor chemistry. Flavor houses, the companies that create artificial flavors, employ chemists who spend their careers synthesizing esters, aldehydes, and other functional-group-bearing molecules and blending them into mixtures that mimic complex natural flavors. Artificial strawberry flavor is not strawberry juice minus the strawberry: it is a carefully balanced mixture of esters, aldehydes, and lactones (cyclic esters), each one contributing a specific olfactory note. The chemist's job is to read the functional group profile of a natural flavor and reproduce it.

A molecule does not get to choose its functional groups any more than a person gets to choose their personality at birth. But once those groups are set, the molecule's behavior follows from them as surely as weather follows from a change in pressure. You can predict, you can anticipate, you can design. That is the power of functional groups.

Try This

Next time you use nail polish remover, read the label. If acetone is listed, sniff carefully. That sweet, sharp smell is a ketone. Then open a bottle of white vinegar and smell that: sharp but different, more acidic, less sweet. That is a carboxylic acid. Two very different personalities, two very different functional groups, and now you know why your nose tells them apart.

Takeaways

• A functional group is a specific cluster of atoms attached to a carbon skeleton that determines how the molecule behaves, smells, dissolves, and reacts.

• The eight major functional groups are: alcohols (-OH), carboxylic acids (-COOH), amines (-NH2), esters (-COO-), ketones (C=O in the middle), aldehydes (-CHO at the end), ethers (-O- bridge), and amides (-CONH-).

• Polarity and hydrogen bonding, both determined largely by functional groups, govern whether a molecule dissolves in water or fat and how easily it evaporates.

• Volatile functional groups (esters, small aldehydes, small amines) are responsible for the smells of fruit, vanilla, cinnamon, and fish.

• Biology uses functional groups as molecular handles: enzymes and receptors recognize specific groups, not whole molecules, and functional groups form the basis of drug design.

• The same carbon skeleton wearing different functional groups becomes an entirely different molecule with entirely different properties, as illustrated by the two-carbon chain that can be ethanol, acetaldehyde, or acetic acid.

• You can now look at a simplified molecule diagram, identify the functional groups present, and make reasonable predictions about whether it will dissolve in water, smell strongly, or interact with biological receptors.

The carbon skeleton is the body. The functional groups are the personality. But there is one more layer to this story, and it is one that took chemists completely by surprise. Two molecules can have identical skeletons and identical functional groups, and still be different. Not subtly different. Mirror-image different. And in biology, that mirror difference is sometimes the difference between a cure and a catastrophe. That is the world of chirality, and it is where we are headed next.

4 Mirror Molecules: Shape and Chirality

Hold up your left hand. Go on, do it right now. Spread your fingers, palm facing you. Now hold up your right hand the same way. They look nearly identical. Same number of fingers, same general architecture, same skin. If you photographed one and placed the photo next to the other, they would appear to be mirror images because they are.

Now try to stack them palm-to-palm so they line up perfectly, every finger matching every finger, every knuckle sitting on its twin. You cannot do it. Your left thumb ends up where your right pinky should be. The two hands are related but irreducibly different. They are, as geometers would say, non-superimposable mirror images.

This is not a hand problem. It turns out to be a molecular problem, a biological problem, and, in one of the darker chapters of twentieth-century medicine, a human catastrophe. The concept is called chirality, from the Greek word for hand. And once you see it, you will never look at a bottle of medication, a sprig of mint, or a lemon the same way again.

What chirality means

The word "chirality" sounds like it belongs in a philosophy seminar, but the idea fits on your kitchen counter. Pick up a mug. Chances are, it has a handle on one side. You can rotate that mug in any direction you like, flip it, tilt it, or spin it, and you will still have the same mug. The handle stays on the handle side. That mug is the same object no matter how you turn it.

Now imagine two mugs that are mirror images of each other, one with the handle on the left and one with the handle on the right. Those are different objects. You cannot rotate one until it becomes the other. A left-handed mug is not a right-handed mug, no matter how long you try.

Molecules work the same way. Most molecules have enough rotational symmetry that spinning or flipping them gets you back to the same molecule. But certain molecules have a geometry that cannot be superimposed on their mirror image, no matter what you do. These are chiral molecules. They come in two versions, left-handed and right-handed, and the chemical community calls those versions enantiomers.

The key structural feature that confers chirality on a carbon atom is deceptively simple: it must be bonded to four different groups. Think of a carbon atom sitting at the center of a tetrahedron, a four-sided geometric solid with one bond pointing to each corner. If all four corners have different passengers, the molecule resembles a glove: you can make a left version and a right version, and they are not the same.

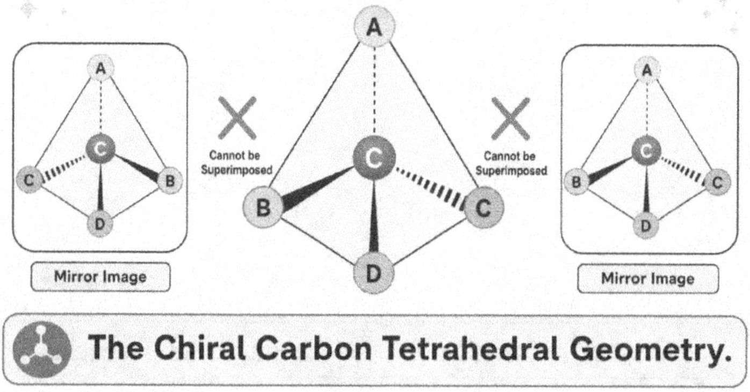

Diagram 4.1: The Chiral Carbon Tetrahedral Geometry

Chemists use the word "stereocenter" or "chiral center" to name that special carbon—one chiral center, one pair of enantiomers: two chiral centers, potentially four versions. And many drug molecules have multiple chiral centers, making it genuinely difficult to deliver the right-handed version to a patient.

Before we go further, pause for a moment and appreciate what is strange here. Two molecules can have the same atoms bonded in the same order. Read their molecular formula, and you get the same answer. Scan the list of what is attached to what, and what is identical. And yet they are different compounds with different properties. The difference lives entirely in three-dimensional space, in the arrangement of atoms around that central carbon. Chemistry is not just about what you have. It is about how it is arranged.

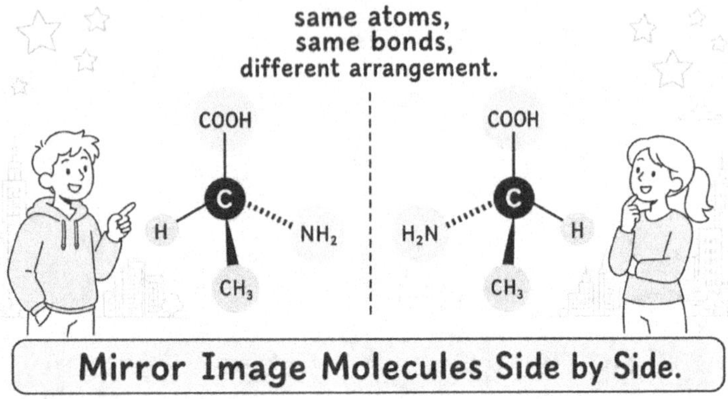

Diagram 4.2: Mirror Image Molecules Side by Side

Enantiomers: same formula, opposite handedness

Here is a good way to think about enantiomers. Imagine two gloves, a left-handed one and a right-handed one. Both are made of the same leather. Both have five fingers. Both have the same stitching pattern. The list of materials is identical. But one fits your left hand, and the other fits your right. They are the same and not the same simultaneously.

Enantiomers carry that same paradox at the molecular level. They have the same molecular formula. They have the same boiling point. They have the same density. They dissolve at the same rate in water. Almost every physical property you can measure is identical between enantiomers. Almost. The one place they differ is in how they interact with other chiral objects. And because biology is built almost entirely of chiral molecules, this difference turns out to be everything.

The classic way chemists distinguish enantiomers in the lab is to shine polarized light through them. Polarized light is light that vibrates in only one plane, like light filtered through a polarized sunglass lens. When polarized light travels through a solution of chiral molecules, the plane of vibration rotates. One enantiomer rotates the plane to the right. Its mirror image rotates it the same number of degrees to the left. This property is called optical activity, and it provides chemists with a way to detect and measure a molecule's handedness.

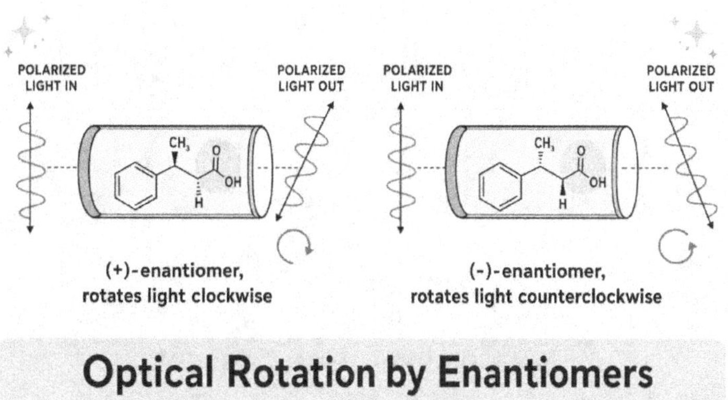

Diagram 4.3: Optical Rotation by Enantiomers

The naming system that chemists use to label enantiomers has changed over time, causing some historical confusion. Older texts sometimes use D and L (for dextro and levo, meaning right and left), a convention inherited from the study of sugars. Newer texts use R and S (from the Latin rectus and sinister, meaning right and left again, just with a fancier assignment rule). For our purposes, the naming conventions are less important than the concept: there are two versions, they are mirror images, and they behave differently inside living things.

One practical detail worth knowing: a mixture of equal parts of the two enantiomers is called a racemate, or a racemic mixture. In a racemate, the optical rotations cancel out, since half the molecules are spinning light one way and the other half are spinning it the other way. For most of pharmaceutical history, drugs were sold as racemic mixtures. This turns out to have been, in at least some famous cases, a serious mistake.

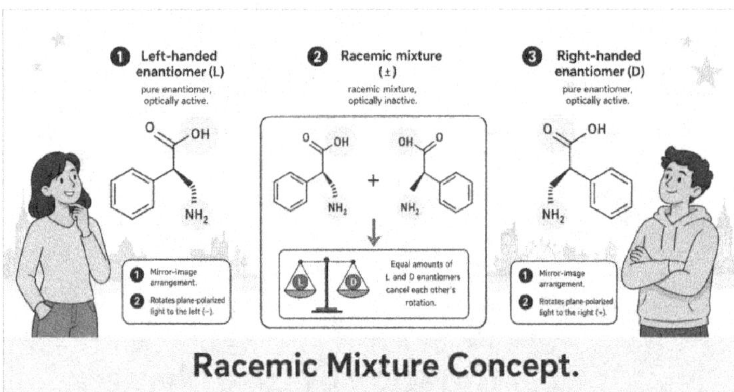

Diagram 4.4: Racemic Mixture Concept

Biological systems are handed too.

Life, as far as we know it on Earth, picked a hand and committed. Every protein in your body is built from amino acids, and nearly all those amino acids are the L-form, the left-handed version. Your cell receptors, your enzymes, your transport proteins, they are all assembled from L-amino acids, which means the three-dimensional architecture of your entire biochemistry is specifically handed. It is as if every lock in your body were made for a left-handed key.

This matters enormously. When a drug molecule arrives at a receptor in your body, it does not just need to have the right atoms. It needs to fit, physically, in three-dimensional space, the way a key fits into a lock. A right-handed key into a left-handed lock will not turn. A molecule of the wrong handedness will either not bind at all, or bind differently, or bind to an entirely different receptor. And all of these outcomes have consequences.

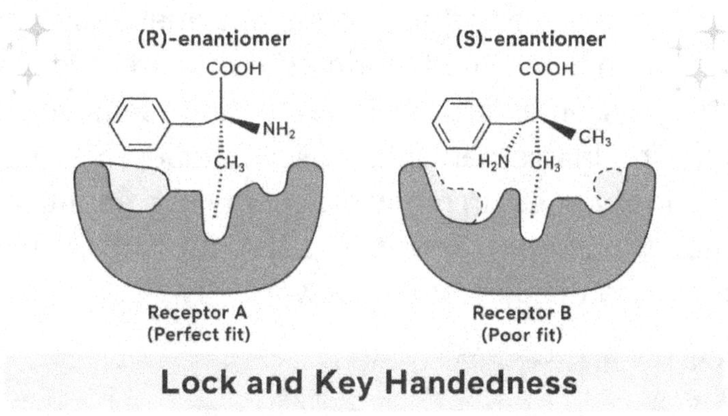

Diagram 4.5: Lock and Key Handedness

The enzyme analogy is equally vivid. An enzyme is a protein that speeds up a chemical reaction by cradling the reacting molecules in a precisely shaped active site. Think of a tailor-made glove that fits one specific hand perfectly. If you try to insert the wrong hand, the glove does not just fit less well; it resists the shape entirely. Enzymes are the same. A molecule with the wrong handedness typically slides right past the active site without reacting.

This selectivity is not accidental. It evolved over billions of years. Life chose L-amino acids, and the machinery built from them is exquisitely tuned to the three-dimensional shapes of the molecules it was designed to handle. The consequence for pharmacology is stark: the handedness of a drug is not an aesthetic preference. It is a functional specification.

Consider ibuprofen, which many people take for headaches or muscle pain. Commercial ibuprofen is sold as a racemic mixture of the S- and R-forms. The S-form is the one that blocks the enzymes responsible for inflammation. The R-form gets converted to the S-form by enzymes in your liver, so the racemic mixture works fine in this particular case. But ibuprofen is one of the lucky ones. In many drugs, the two enantiomers do not interconvert. They stay separate. And the wrong one can do things you do not want.

The thalidomide story

In the late 1950s, a West German pharmaceutical company, Chemie Grunenthal, introduced a new sedative called thalidomide. It was marketed as a safe sleeping aid and an anti-nausea drug, and it was promoted specifically for use by pregnant women suffering from morning sickness. By 1960, it was available in forty-six countries under various brand names. In the United Kingdom alone, it was being sold over the counter without a prescription.

Thalidomide was, in one sense, effective. It did reduce nausea. And in short-term studies in adults, it appeared remarkably non-toxic. The company pointed to the low acute toxicity as evidence of safety. The U.S. Food and Drug Administration reviewer Frances Kelsey, who would later receive a Presidential award for her work, was skeptical. She kept asking for better data on long-term effects and refused to approve it for American markets. That skepticism turned out to be one of the most consequential acts of scientific caution in pharmaceutical history.

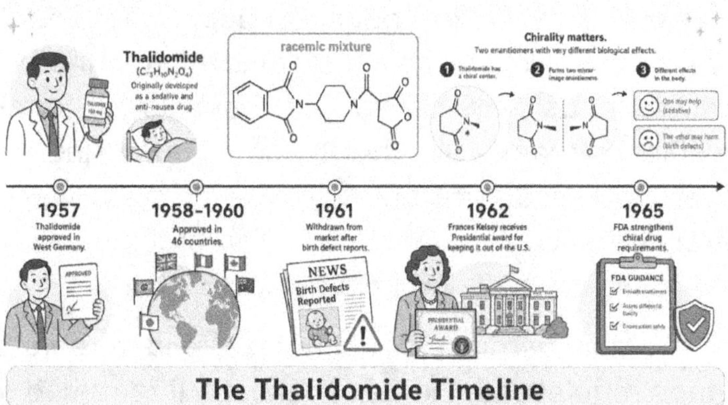

Diagram 4.6: The Thalidomide Timeline

Between 1957 and 1961, approximately ten thousand children were born with severe limb defects, a condition called phocomelia, to mothers who had taken thalidomide during early pregnancy. The drug interfered with the development of limbs in the fetus. It was withdrawn from most markets in 1961, after researchers in Germany and Australia independently connected the defects to the drug. The damage had already been done.

Here is where the connection to chirality comes in. Thalidomide is a chiral molecule. It has two enantiomers. The R-form provides the sedative and anti-nausea effects that make the drug seem useful. The S-form is teratogenic, meaning it disrupts fetal development. Thalidomide was sold as a racemic mixture, so every dose contained both enantiomers.

There is a bitter addendum to this. Even if pharmaceutical chemists in the 1950s had been able to separate only the R-form and administer that alone, it would not have been enough. Thalidomide undergoes a process called racemization in the body: it spontaneously interconverts between its two forms in the bloodstream. So the R-form, once ingested, generates the S-form anyway. The problem was not merely the racemic formulation. The problem was the drug itself, in a body with chiral biochemistry.

The thalidomide disaster did not just end careers and change regulations, though it did both. It reframed how the entire pharmaceutical industry thought about chirality. Before 1960, most chemists considered the two enantiomers of a drug to be essentially the same compound. After thalidomide, that assumption was finished. The handedness of a molecule was now understood to be a first-order safety consideration, not an afterthought.

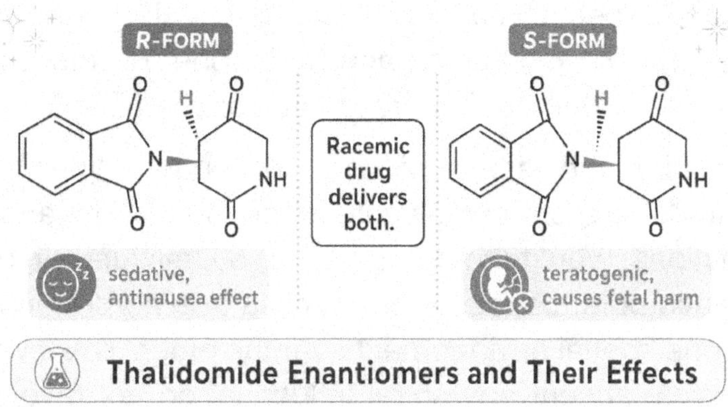

Diagram 4.7: Thalidomide Enantiomers and Their Effects

The story does not end in tragedy, however. Thalidomide was eventually reapproved, decades later, for a narrow set of uses: a type of leprosy complication called erythema nodosum leprosum, and later certain forms of blood cancer. It is a genuinely useful drug against those conditions. But it is now dispensed under some of the strictest prescribing controls in pharmaceutical history, with mandatory pregnancy prevention programs for any patient who might conceive. A terrible drug, in the wrong hands, in the wrong era, became a cautionary tale that made medicine safer.

How modern drugs solve the handedness problem

The pharmaceutical industry today treats chirality as a central engineering challenge, not an edge case. Developing a single pure enantiomer of a drug, rather than a racemic mixture, is called chiral synthesis or asymmetric synthesis, and it is one of the more technically demanding areas in all of chemistry.

Consider the scale of the challenge. When you synthesize a chiral molecule in the lab using standard reactions, you tend to get a 50/50 mixture of both enantiomers, because the reaction has no preference for one spatial arrangement over the other. The laws of physics, absent any chiral influence, do not favor left over right. Getting only one enantiomer requires either separating the mixture after the fact or incorporating a chiral influence into the reaction so that the desired enantiomer forms preferentially.

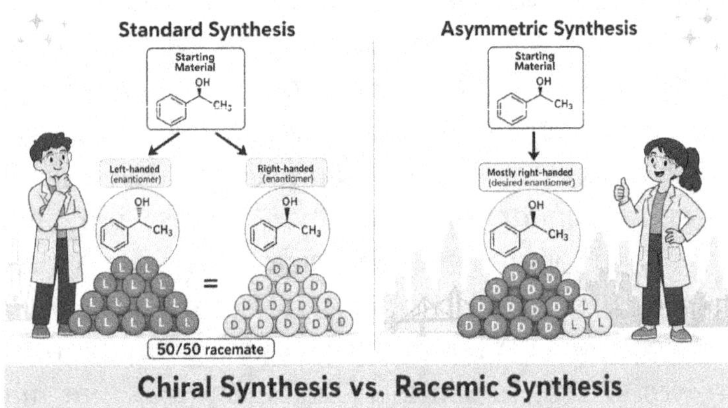

Diagram 4.8: Chiral Synthesis vs. Racemic Synthesis

Separating enantiomers after synthesis, a process called chiral resolution, is often achieved by exploiting biology's own handedness. You can form a salt or a complex between your racemic mixture and a pure chiral molecule, typically one taken from nature, and the two resulting compounds (left-drug paired with right-resolver, and right-drug paired with right-resolver) become diastereomers. Diastereomers are not mirror images, so they have different physical properties and can be separated by standard methods such as crystallization. It is an elegant trick, using chirality to sort chirality.

The alternative approach, asymmetric synthesis, is chemically more satisfying. Here, chemists design reactions that use a chiral catalyst or chiral starting material to ensure that the new bond forms preferentially on one face of the molecule. The catalyst acts like a chiral mold, shaping the product as it forms. The 2001 Nobel Prize in Chemistry was awarded to William Knowles, Ryoji Noyori, and Barry Sharpless for their development of asymmetric catalysis reactions. Those reactions now underpin the manufacturing of dozens of drugs, including L-DOPA for Parkinson's disease and many antibiotics.

Getting chirality right in a drug costs money, substantial amounts of it. Developing an enantiopure drug (one that contains only the desired single enantiomer) can add hundreds of millions of dollars to the development cost of a new compound. Regulatory agencies worldwide now routinely require pharmaceutical companies to test both enantiomers of any chiral drug separately before approval and to justify the formulation choice. The lesson of thalidomide was expensive to learn, but it has been thoroughly learned.

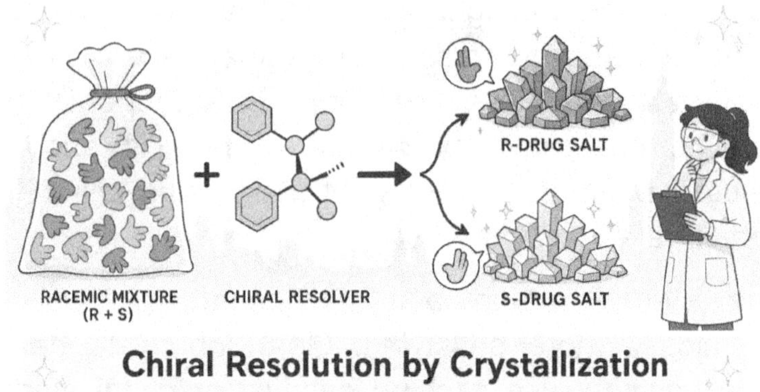

Diagram 4.9: Chiral Resolution by Crystallization

There is a concept called "chiral switching" worth knowing about. When the patent on a racemic drug expires, and the drug becomes generic, the original manufacturer sometimes responds by developing and patenting the single active enantiomer as a new drug. Omeprazole, sold for years as the racemic drug Prilosec for acid reflux, was followed by esomeprazole, sold as Nexium, which is just the S-enantiomer. Nexium is genuinely slightly more effective. The business motivation to file a new patent is obvious. The practice is controversial, but it does sometimes produce a clinically improved product.

Smell, taste, and chirality in everyday life

All of that heavy pharmacology is building toward something surprisingly pleasant: the reason your toothpaste smells like toothpaste and not like rye bread.

Carvone is a molecule with one chiral center. It exists as two enantiomers. Both are colorless liquids, both have the same molecular formula, and both have the same boiling point. If you poured them into two unlabeled vials and ran every standard chemical analysis on them, you would struggle to tell them apart.

Then you would smell them.

(R)-Carvone smells exactly like spearmint. It is the dominant odor compound in spearmint oil, the one responsible for the clean, bright, slightly sweet smell universally associated with toothpaste, chewing gum, and mint tea. (S)-Carvone smells like caraway seeds, the warm, slightly bitter, earthy spice found in rye bread and certain European liqueurs. Put both vials side by side and smell them in turn, and you will have, in that moment, the most visceral possible demonstration that molecular handedness is real, consequential, and genuinely strange.

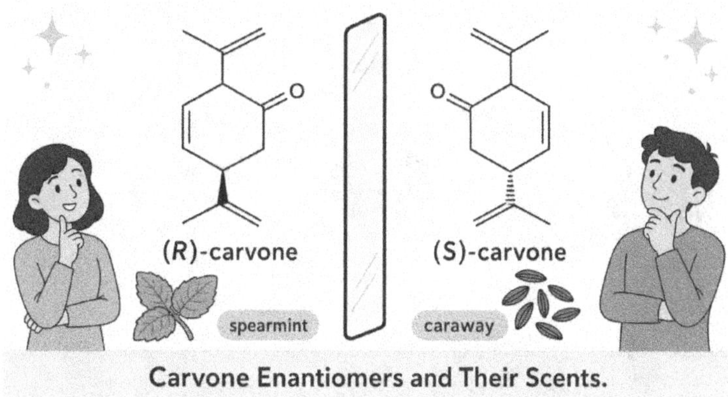

Diagram 4.10: Carvone Enantiomers and Their Scents

Your olfactory receptors, the proteins in your nose that detect smell molecules, are themselves chiral. They are made of amino acids, remember, which are L-form. So your smell receptors fit one enantiomer of carvone perfectly, and the other less well, activating different receptor proteins and sending different signals to your brain. When you smell spearmint, you are running a chirality experiment. Your nose is doing analytical chemistry that separates enantiomers more reliably than most lab equipment.

Limonene is another example that most people encounter daily. It is the dominant fragrance compound in citrus peels. (R)-limonene, the right-handed version, is what makes oranges smell like oranges, that bright, fruity, slightly sweet citrus blast. (S)-limonene, the left-handed version, is found in lemon peel and has a sharper, more piney quality. Same atoms, same bonds, mirror-image arrangement. Different smell. Orange versus lemon, built from a single chiral center.

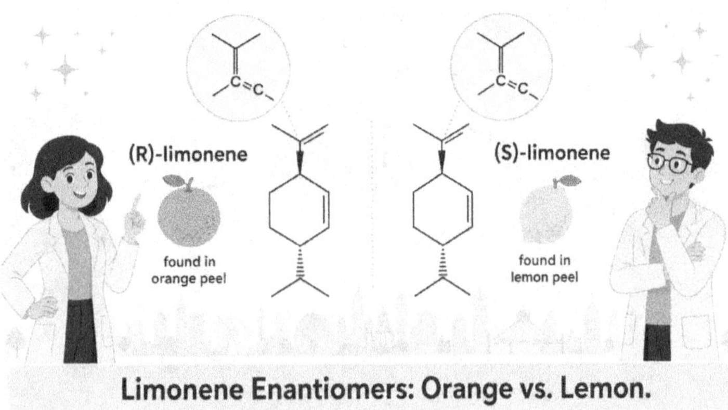

Diagram 4.11: Limonene Enantiomers: Orange vs. Lemon

The flavor world is equally full of chiral surprises. Aspartame, the sweetener in many diet sodas, is chiral. Only one form tastes sweet; the other is essentially tasteless. Amino acids themselves are chiral, and it turns out that L-amino acids, the ones your body uses to build proteins, generally taste different from D-amino acids. Some D-amino acids taste sweet, which is why research on synthetic sweeteners pays close attention to chirality. The world of taste is a world of handedness.

Even some drugs that have nothing to do with chirality as a design feature end up being affected by it through metabolism. The body processes each enantiomer through different enzymatic pathways, which means one form might be cleared from the blood more quickly, converted to a different compound, or accumulate in tissues at a different rate—the hands of a molecule echo through an entire biological system.

Diagram 4.12: Chirality in Everyday Products

Try this

Go to your kitchen and find something with spearmint flavoring, a stick of gum, a toothpaste, or a mug of mint tea. Smell it. Then, if you have rye bread or caraway seeds, smell those too. You are experiencing the same molecular formula expressing two completely different identities. You are, without any lab equipment, detecting a difference that comes down to the three-dimensional arrangement of atoms around a single carbon atom. That is chirality, live and in person.

Takeaways

• You can explain chirality using the glove-and-hand analogy: molecules with four different groups around a carbon atom exist in two non-superimposable mirror-image forms called enantiomers.

• You can describe why enantiomers have identical melting points and formulas yet behave completely differently inside living things, because biological receptors and enzymes are themselves chiral.

• You can explain the thalidomide disaster: one enantiomer had therapeutic value, the other caused fetal harm, and the racemic mixture delivered both, with irreversible consequences for thousands of families.

• You can articulate why modern pharmaceutical development treats chirality as a core engineering problem, not an afterthought, and why single-enantiomer drugs often represent genuine clinical improvements.

• You can account for why spearmint smells like spearmint and caraway smells like caraway: they are enantiomers of carvone, and your olfactory receptors are handed too.

• You can recognize chirality as a pattern that runs through medicine, flavor, fragrance, and the entire biochemistry of life, not a laboratory curiosity but a property woven into the fabric of how biology works.

• You can read a news story about a drug recall or a reformulation as a single enantiomer and understand, at the molecular level, exactly what problem the scientists were solving.

The molecular map has a geometry, and now you can read it. In the next chapter, we turn from the shape of molecules to what they do: chemical reactions, told not as equations but as stories of bonds breaking and forming, energy released and absorbed, molecules transforming into entirely different things. The language you have been building, chapter by chapter, has always pointed toward this. Time to watch the molecules move.

5 The Art of the Deal: How Molecules React

You've done it a hundred times without thinking. You pull a box of baking soda from the cabinet, pour a splash of white vinegar into a bowl, and add a spoonful of the white powder. The mixture roars to life. It hisses and foams, climbs toward the rim, and spills over the edge if you're not careful. Kids love it. Adults use it to clean drains. But here's the thing no one told you in middle school: what you just witnessed is not a trick, not a special effect, and not really a "reaction" in the vague hand-wavy sense that word usually gets used. You just watched a negotiation play out at a scale you cannot see, conducted by particles moving faster than sound, striking each other with enough force to tear themselves apart and rebuild into something completely new.

A gas escaped from that bowl. Carbon dioxide, the same stuff you exhale, the same stuff that bubbles in your soda. It wasn't in the vinegar, and it wasn't in the baking soda. It was built, atom by atom, in the span of a few seconds, out of pieces that used to belong to entirely different molecules. New bonds formed. Old bonds snapped. Two substances went in. Three substances came out. That's a reaction. Not a symbol on a chalkboard. A genuine molecular transformation.

This chapter is about how that works. And not just the vinegar-and-baking-soda version. Every drug your body metabolizes, every color that appears when you sear a steak, every plastic that hardens in a mold, every breath you take: they all run on the same basic machinery. Molecules meet. Bonds break. New bonds form. Something entirely new exists where nothing existed before. Once you see the pattern, you'll see it everywhere.

What a reaction really is: bonds

Let's start with the molecular truth, because it's both simpler and stranger than chemistry class ever made it sound.

You already know from earlier chapters that bonds hold molecules together. A bond is a shared agreement between two atoms: we will both donate an electron, share the resulting pair, and, in exchange, both of us become more stable. That agreement is not permanent. Under the right circumstances, it can be renegotiated. One party can leave. A new partner can arrive. The bond between two atoms can break, and a bond to a completely different atom can form. That, in its entirety, is what a chemical reaction is. Nothing more. The rearrangement of who is bonded to whom.

Think of it like a neighborhood of houses connected by roads. A reaction is what happens when some of those roads close and new roads open. The houses themselves (the atoms) don't change. Carbon is still carbon. Oxygen is still oxygen. But the network connecting them rewires. What was a dead end becomes a through street. What was a busy junction gets shut down. When you zoom out and look at the new map, the neighborhood looks completely different, even though every house is still standing.

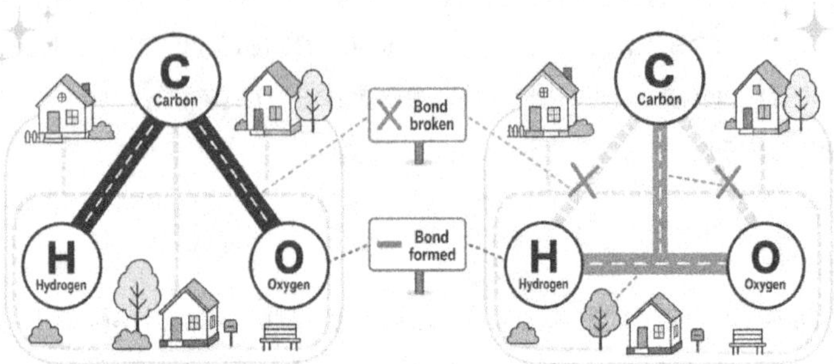

Bonds as roads in a molecular neighborhood

Diagram 5.1: Bonds as roads in a molecular neighborhood

The vinegar (acetic acid) and baking soda (sodium bicarbonate) story plays out this way. In acetic acid, a hydrogen atom is bonded to an oxygen atom, and that particular bond is unusually willing to release its hydrogen. In sodium bicarbonate, there's a carbon atom bonded to three oxygens, a configuration that, when pushed, can give up a piece of itself as carbon dioxide gas. When these molecules meet, and the hydrogen from the acid finds a new bonding partner among the bicarbonate's oxygens, the reshuffling cascades: the bicarbonate falls apart, the oxygen gets its new partner, and the carbon that used to be in the solid bicarbonate drifts away, bonded to two oxygens as a gas. The hissing you hear is literally millions of carbon dioxide molecules escaping the liquid all at once.

What drives this? Why do the atoms want to reshuffle at all? There's an energy accounting happening beneath the surface. The new arrangement of bonds in the products is lower in energy than the old arrangement in the reactants. Nature, at the molecular level, is an enthusiastic minimizer. Given the chance to drop to a lower energy state, molecules will take it. The reaction releases some of that energy difference as heat and kinetic energy (which is why many reactions warm up their surroundings). The fizzing bowl of vinegar and baking soda is, at its heart, a controlled release of energy, expressed as molecular reorganization.

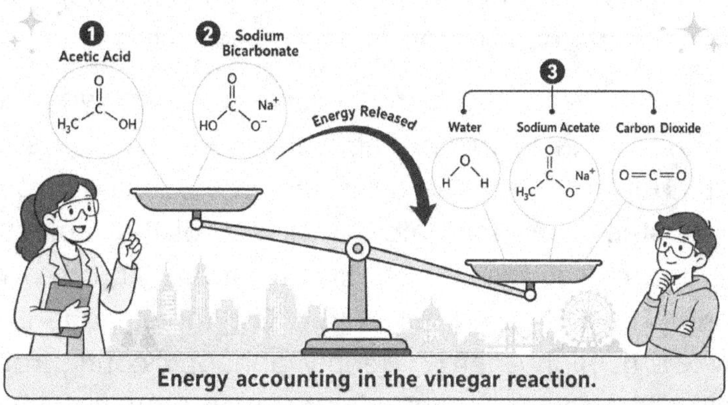

Diagram 5.2: Energy accounting in the vinegar reaction

But here's the part that makes chemists' eyes light up: if rearranging bonds releases energy and nature always seeks the lowest energy, why doesn't everything react with everything all the time? Why does your bottle of acetic acid sit peacefully on the shelf for months without decomposing into something else?

Why doesn't the baking soda in your cabinet spontaneously transform the moment air touches it?

Because getting from the old arrangement to the new one isn't free, there's a hill in the way.

Activation energy: why molecules need a push

Imagine you're trying to push a boulder over a hill. The far side of the hill is lower than where you started, so if you get the boulder to the top, it'll roll down the other side on its own and end up in a lower, more stable position. Great. But first, you have to push it to the top. That push requires energy, regardless of what happens afterward. You can't skip the climb.

Molecules face the same problem. Even when the product state is more stable than the starting state, getting from one to the other requires passing through an unstable intermediate. At the top of the molecular hill, the old bonds are starting to break, and the new bonds haven't fully formed yet. The molecule is in an awkward, high-energy, deeply unhappy configuration called the transition state. It exists for an unimaginably brief moment, perhaps a trillionth of a second, before either collapsing back into reactants (the boulder rolls back) or completing the conversion into products (the boulder crests and rolls forward).

The energy required to reach the top of this hill is called the activation energy. It's the price of admission. Every reaction has one. Some hills are tiny, barely a speed bump, and those reactions zip along at room temperature without any encouragement. Others are enormous, and those reactions won't happen unless you supply a significant energy jolt, usually in the form of heat.

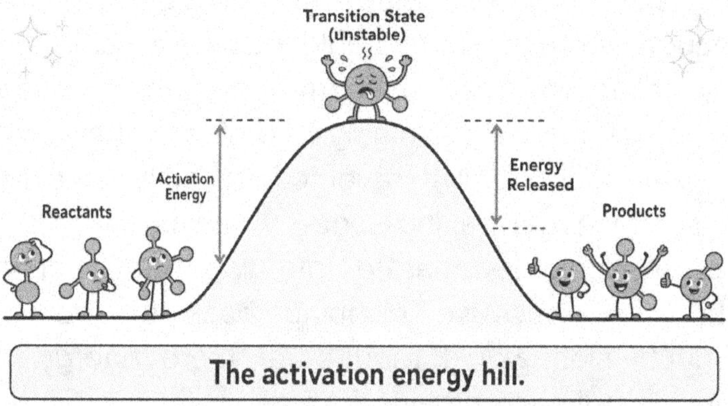

Diagram 5.3: The activation energy hill

This is why you can leave a piece of butter on the counter at room temperature without it bursting into flame, even though butter combusting in air would release a spectacular amount of energy and end up at a much lower, more stable energy state. The activation energy for combustion is enormous. At room temperature, butter molecules lack sufficient thermal energy to reach the transition state. Put butter in a pan on high heat, though, and you pump in enough energy to get things started. It browns first (the Maillard reaction, a set of bond-breaking and forming events between sugars and proteins), and if the temperature gets extreme enough, it burns. The energy was always available to release. The activation energy was the gatekeeper keeping it locked up.

Here's a beautiful subtlety: activation energy also explains why reactions are reversible in principle but not always in practice. If you can push the boulder up from the left, you can theoretically push it up from the right. But if the right-side slope is very steep (the products are much more stable than the reactants), the boulder would need an enormous push to go back. Most reactions that are highly favorable in one direction are extremely difficult to reverse for exactly this reason. Your vinegar-baking-soda mixture doesn't spontaneously regenerate into two neat, separate substances because climbing back up from the products side would require far more energy than nature is willing to provide for free.

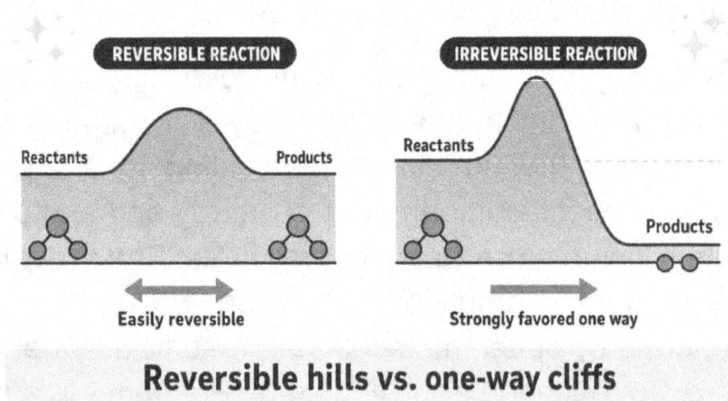

Diagram 5.4: Reversible hills vs. one-way cliffs

Temperature, by the way, is just a measure of how fast molecules are moving. A higher temperature means molecules are bouncing around with more kinetic energy, colliding harder, and more of them have enough energy to reach the activation energy threshold at any given moment. This is why heating speeds up reactions: you're not changing the hill, you're just giving more molecules enough energy to climb it.

Catalysts as molecular matchmakers

Now here is one of the most elegant ideas in all of science. What if you could lower the hill without adding heat? What if you could reshape the energy landscape so that the same transition from reactants to products required less energy to achieve?

You can. And the thing that does it is called a catalyst.

A catalyst is a substance that provides an alternative pathway for a reaction, one with a lower activation energy, and then gets itself back out of the deal unchanged at the end. It participates without being consumed. It makes the deal easier to close without taking a cut.

Picture the catalyst as a mutual friend who introduces two shy people at a party. Without the mutual friend, the two of them might stand on opposite sides of the room all night, each waiting for the other to make a move, the activation energy of introducing themselves being just too high to overcome. The mutual friend walks over, makes the introduction, gets things started, and then drifts back to the buffet table. The conversation (the reaction) proceeds on its own. The mutual friend is the same person they were before the introduction. They've enabled something without being part of its outcome.

The catalyst as party matchmaker.

Diagram 5.5: The catalyst as party matchmaker

Chemists use catalysts constantly. The platinum in your car's catalytic converter brings together unburned hydrocarbons and oxygen with enough intimacy for them to react efficiently, converting toxic carbon monoxide into less harmful carbon dioxide at temperatures achievable inside an exhaust pipe. The platinum itself emerges from those reactions unchanged, ready to facilitate the next round.

Industrial catalysts are similarly transformative. The Haber process, which produces the ammonia that grows roughly half of the world's food, depends on an iron catalyst to bring nitrogen and hydrogen gases close enough and in the right orientation for their bonds to break and reform into ammonia. Without the catalyst, the reaction would require temperatures and pressures so extreme they'd be industrially impractical. Iron lowers the hill. Billions of people eat, in part, because of iron doing its matchmaking job inside giant steel reactors.

The key insight is that catalysts are not magic. They don't invent new energy or violate the rules of thermodynamics. The overall energy accounting, the difference between reactant energy and product energy, stays the same. They find a cleverer route over the mountain, a pass through the peaks rather than the sheer face.

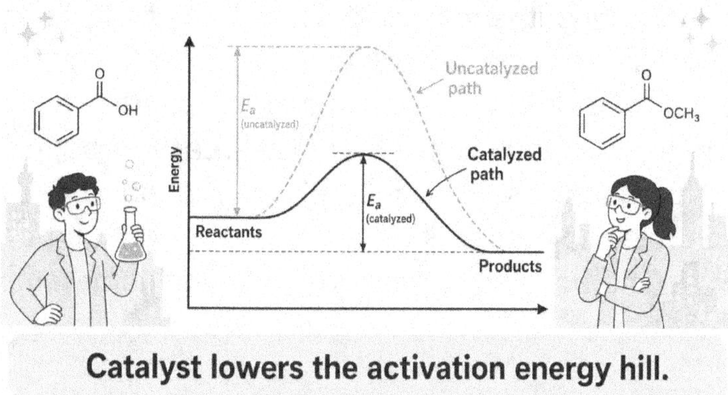

Diagram 5.6: Catalyst lowers the activation energy hill

Enzymes: biology's catalysts

If industrial catalysts are impressive, enzymes will make you feel that evolution has been quietly outdoing our best engineers for three and a half billion years.

An enzyme is a biological catalyst. It's a protein, a large and exquisitely folded molecule, that has evolved to do one thing with breathtaking precision: bring a specific molecule (or pair of molecules) into exactly the right orientation for a specific bond-breaking or bond-forming event to occur, dramatically lower the activation energy, and then release the product and reset for the next round. Over and over. Thousands of times per second.

The image you might have encountered is the lock-and-key model. The enzyme has a pocket, called the active site, that is shaped to accept one particular type of molecule (the substrate) like a key fitting a lock. When the right substrate docks into the active site, the enzyme grips it, sometimes slightly strains it, nudges its bonds into a configuration they weren't quite in before, and dramatically reduces the energy required to reach the transition state. The reaction happens. The product, now a slightly different shape from the substrate, no longer fits the active site well and drifts away. The enzyme is back to its original form, empty, ready.

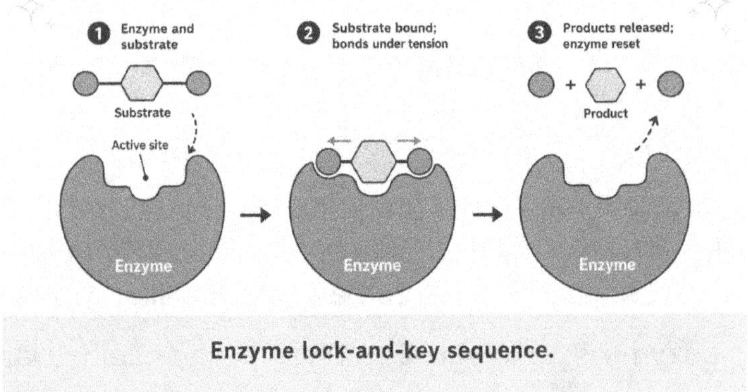

Diagram 5.7: Enzyme lock-and-key sequence

Here's the analogy that really nails it: enzymes are speed-dating facilitators. In a regular social setting, two people might wander a party for an hour before ending up in a conversation that could have happened in the first five minutes. Speed dating eliminates that friction. Each person gets seated; their partner rotates in; the two interact in a structured context for three minutes, and then the next partner arrives. The facilitator designs the structure. The reactions (hopefully) happen efficiently. The facilitator doesn't go home with anyone.

Your body runs on enzymes. A specific enzyme catalyzes every step in digesting the food you ate for breakfast. The enzyme amylase in your saliva starts breaking the bonds in starch molecules the moment food enters your mouth. In your stomach, pepsin is tearing apart protein bonds. In your small intestine, a whole committee of enzymes is handling sugars, fats, and proteins simultaneously. Each enzyme knows its substrate, fits it precisely, lowers the activation energy for exactly one type of bond event, and produces exactly one type of output.

This specificity is what makes enzyme biology so extraordinary. A general-purpose acid might break many different bonds indiscriminately. An enzyme breaks one bond in one molecule in one specific orientation and leaves everything else alone. That is why drugs that target enzymes are so powerful, and why designing them requires an understanding of the molecular geometry of the active site with near-atomic precision. We'll come back to this idea in Chapter 6.

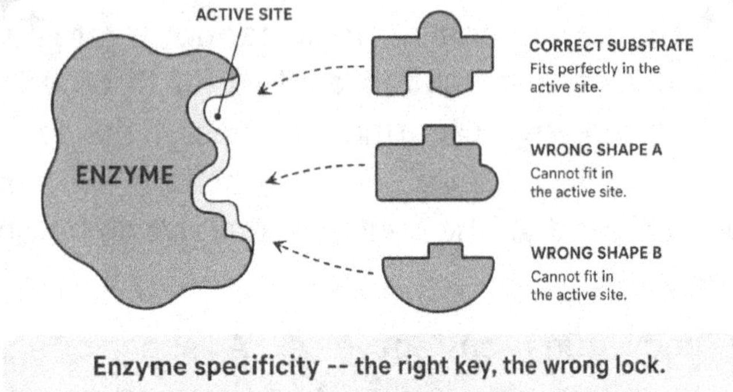

Diagram 5.8: Enzyme specificity, the right key, the wrong lock

One more layer of sophistication: enzymes themselves are regulated. The body doesn't want every enzyme running at full speed all the time. Some enzymes can be switched on or off by the binding of a small molecule at a separate site, not the active site, that changes the enzyme's shape just enough to make the active site less (or more) welcoming. This is called allosteric regulation. It's how the body turns metabolic pathways up or down in response to changing needs. Hungry? Certain enzymes switch on to release stored energy. Full? Others switch on to store the surplus. The whole system operates without a central controller, coordinated entirely by molecular shape changes rippling through networks of enzyme interactions.

Substitution, addition, and elimination

So far, we've talked about reactions in general terms. Now let's meet the plot types. Just as every story in literature can be loosely sorted into a handful of narrative arcs, every reaction in organic chemistry can be sorted into a handful of story types based on what changes and how.

Three of the most important are substitution, addition, and elimination. You don't need to memorize them as category labels. You need to recognize the narrative pattern because once you do, you'll see that chemistry is telling variations of the same three stories over and over.

Substitution: the swap

In a substitution reaction, one atom or group on a molecule gets replaced by a different atom or group. The core structure of the molecule stays intact. Only one player leaves, one player arrives, and they trade places.

Think of it like a game of musical chairs where one person is already seated and a second person pushes them out and takes the chair. The chair (the carbon skeleton) doesn't go anywhere. The person who was sitting (the leaving group) walks away. The newcomer sits down. The room's basic furniture remains unchanged; the occupants have shifted.

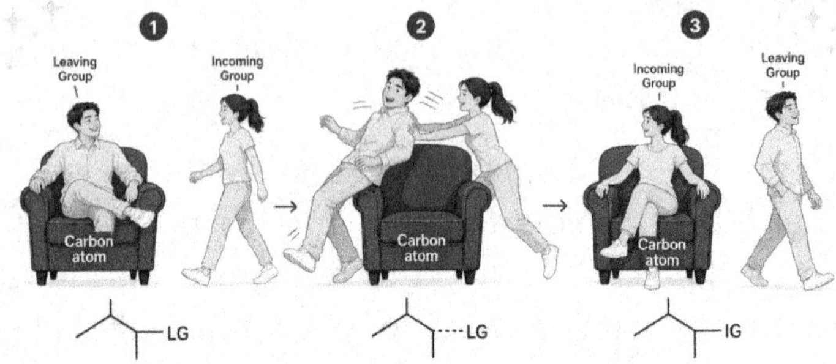

Substitution as musical chairs.

Diagram 5.9: Substitution as musical chairs

Substitution reactions are everywhere in organic chemistry, but you encounter a beautifully direct example in how your body handles certain fat-soluble compounds. The liver essentially performs substitution chemistry by replacing a relatively insoluble organic compound with a water-soluble group, making the compound easier to excrete. The core molecular structure barely changes. The swap makes all the difference.

In addition, the merger

In an addition reaction, two separate molecules join together to become one. Nothing leaves. A bond opens up on one molecule, and parts of a second molecule attach at the opened ends. What was a double bond (two atoms sharing two pairs of electrons) becomes a single bond with new attachments.

Imagine two puzzle pieces lying separately on a table. One piece has an obvious open notch in the middle. The other piece breaks into two halves, and each half snaps into one end of the notch. Now they're all one connected piece. No piece was left on the table. The total inventory of atoms is identical; they're just joined together differently.

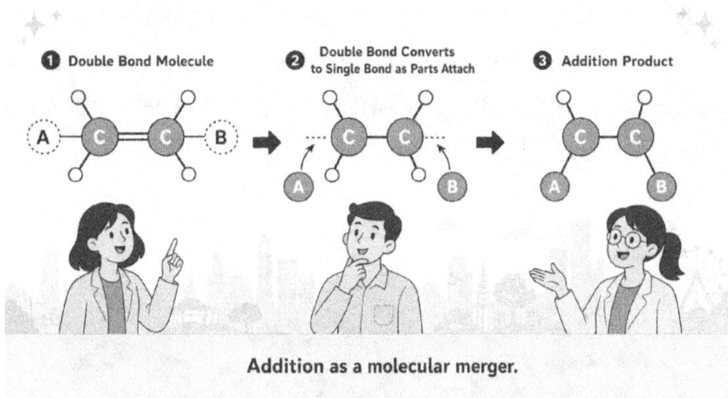

Diagram 5.10: Addition as a molecular merger

The hydrogenation of vegetable oils is a real-world addition story. Liquid vegetable oils contain molecules with double bonds between some of their carbons. Adding hydrogen across those double bonds, with a metal catalyst doing the facilitating, converts double bonds to single bonds and turns liquid oil into solid or semi-solid fat. That's how margarine is made. The transformation in texture is a direct consequence of addition reactions that change the geometry and flexibility of fat molecules.

Elimination: the departure

Elimination is the opposite of addition. Where addition was a joining, elimination is a departure. A molecule loses a small fragment (often two atoms from adjacent positions), a bond breaks, and in compensation, a new double bond forms between the atoms that lost their attachments.

Picture a couple who were separately holding hands with their neighbors in a long chain. They let go of their neighbors and clasp hands with each other instead. The neighbors are now free (they've been eliminated as attached groups), and the couple in the middle has formed a new, closer bond.

Elimination as a departure and reconnection.

Diagram 5.11: Elimination as a departure and reconnection

Elimination reactions are of enormous importance in industry and biology. When polymers are synthesized, when your body degrades certain compounds, when alcohol is converted to something more reactive, elimination reactions are often the key step. The beautiful thing is that the same three patterns, swap, join, and depart, appear whether you're talking about tiny molecules or enormous polymer chains. Chemistry's vocabulary is relatively small. Its sentences are endless.

What controls reaction speed

You now understand what a reaction is (bond-breaking and forming), why it needs a push (activation energy), how catalysts lower that push, and what story types the rearrangement can follow. The final question is: once you've got a reaction that can proceed, what determines how fast it actually goes?

The answer turns out to be wonderfully intuitive. Three main factors control reaction rate, and all three come down to the same underlying principle: the more often molecules collide with sufficient energy and in the right orientation, the faster the reaction proceeds.

Temperature

We touched on this briefly, but it's worth making explicit. Temperature is molecular speed. A hot sample of molecules has more molecules moving fast, which means more collisions per second, and more of those collisions carry enough energy to clear the activation energy barrier. Double the reaction rate roughly every ten degrees Celsius, as a rule of thumb, and you've got a feel for how powerfully temperature controls reaction speed.

This is why your refrigerator works. Bacteria on your food metabolize (carry out chemical reactions) to reproduce and grow. Lower the temperature, and their molecular machinery slows dramatically. Reactions that would happen in hours at room temperature take days in the cold. Your milk lasts longer in the fridge, not because bacteria disappear, but because their enzymes slow to a crawl. Freeze the milk, and you slow things so far that reactions take months. The same chemistry, at three different temperatures, produces three wildly different outcomes.

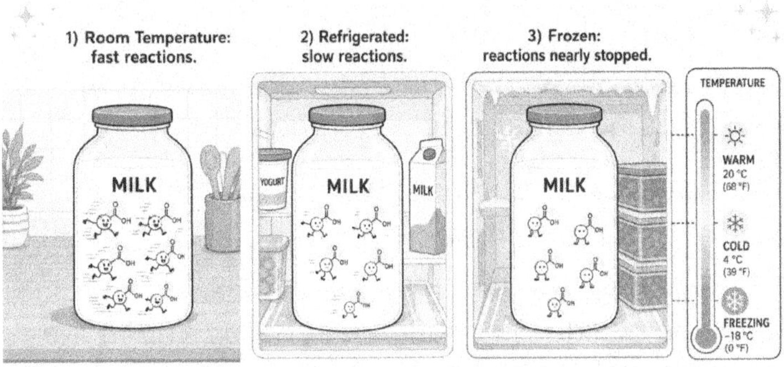

Temperature and reaction speed in the kitchen.

Diagram 5.12: Temperature and reaction speed in the kitchen

Concentration

If temperature is about how energetic the collisions are, concentration is about how often collisions happen in the first place. More molecules in the same space means more crowding, more encounters, more opportunities for the right two molecules to meet at the right moment. Reactions generally speed up when you increase the concentration of the reacting molecules.

This is the principle at work when you increase a drug's dose. More drug molecules in your bloodstream means more frequent encounters with the enzyme or receptor that the drug is designed to interact with. There's a limit (the enzyme or receptor gets saturated), but within the working range, concentration and speed go hand in hand.

It's also why reactions slow down as they proceed. As reactants are consumed, their concentrations fall, collisions become less frequent, and the reaction rate slows. That early fizzing explosion from your vinegar and baking soda slows to a gentle hiss and then stops, not because the reaction has changed, but because the reactants have run out. The hill didn't get higher. The crowd of molecules climbing it just thinned out.

Surface area

This one surprises people. For reactions involving a solid, only the molecules on the surface of the solid are accessible to react with whatever is in the surrounding liquid or gas. If you want a faster reaction, you need to expose more surface.

Crushing a solid into powder dramatically increases its surface area. A single sugar cube has perhaps a few square centimeters of exposed surface. Grind that same sugar into fine powder, and you have thousands of times more surface exposed to the air and any reacting agent. Powdered materials often react explosively, while the same material in bulk form would react slowly or not at all. Flour dust in a grain elevator is genuinely explosive. A loaf of bread is not. Same molecules. Different surface area.

This is also why antacid tablets work faster when you chew them before swallowing: you've just dramatically increased the surface area of the calcium carbonate available to react with stomach acid—chemistry, not magic. The same principle operates in catalysis: platinum catalysts in catalytic converters are often prepared as fine powders or porous structures to maximize the surface area available for contact with exhaust molecules.

Try this

Next time you make tea or coffee, brew a cup with room-temperature water and see how long it takes. Then brew the same amount with hot water. You're watching activation energy in real time: the same reactions that extract flavor and caffeine from the leaves or grounds proceed maybe fifty times faster at near-boiling temperatures than at room temperature. The molecules in hot water have far more energy. More of them can clear the activation energy barrier on any given collision. The hot cup is ready in minutes. The cold one takes hours. Same chemistry, different energy.

Takeaways

• You can explain a chemical reaction at the molecular level: bonds break, and bonds form, but atoms themselves are permanent and unchanged.

• You can describe how energy is released or absorbed in a reaction: new bonds in the products are either lower or higher in energy than the old bonds in the reactants, and the difference is what the reaction releases or stores.

• You can name and explain activation energy as the hill every reaction must climb before it can roll downhill.

• You can describe what a catalyst does: it lowers the activation energy by providing an easier route, while participating without being consumed.

• You can explain why enzymes are extraordinary catalysts, each shaped to accept one substrate, lower the energy barrier for one bond event, and release the product to start again.

• You can recognize the three story types of organic reactions, substitution, addition, and elimination, as a swap, a merger, and a departure, and you can spot them in your body and your kitchen.

• You can predict how temperature, concentration, and surface area will speed up or slow down a reaction, because you understand how each one changes how often and how energetically molecules meet.

Now that you can see the machinery of transformation, bonds breaking, hills being climbed, enzymes orchestrating the whole ballet, it's time to ask a more pointed question. What happens when humans deliberately design molecules to interfere with that machinery? When a chemist sits down and says: I want to block that enzyme, or trigger that reaction, or mimic that natural compound closely enough to fool the body's receptors? That is the story of drugs. And it starts exactly where you'd expect: at the molecular level, with shape, energy, and a very well-designed key seeking the exact right lock.

6 Designed to Heal: How Drugs Are Built

Open your medicine cabinet. Go ahead, I'll wait.

Somewhere in there, you probably have aspirin, ibuprofen, and acetaminophen. Maybe an antihistamine for allergy season. A tube of antibacterial. If you look at those bottles the way most people do, you see vague shapes and brand names. The labels promise relief and list a string of warnings in six-point font that nobody reads.

But here is what you are actually looking at: a collection of precisely shaped carbon molecules, each one engineered to find a single specific protein somewhere inside your body, latch onto it like a key sliding into a lock, and trigger exactly one change. Not a general change. Not a "feel better" wash of chemistry. One specific molecular event. And that event, multiplied across billions of cells, is what makes your headache go away.

Those three pain relievers in your cabinet, aspirin, ibuprofen, and acetaminophen, are not interchangeable. They have different shapes. They hit different targets. They work by different mechanisms. And the fact that all three knock out a headache is almost accidental, a convergence of different molecular strategies arriving at the same practical outcome. Understanding that is understanding drug design at its core.

This chapter is about the science of making molecules that heal. It is one of the most practically important applications of organic chemistry and one of the most expensive and difficult things human beings have ever attempted. The average drug that reaches a pharmacy shelf has survived a journey that took roughly a decade, consumed somewhere between one and three billion dollars, and outlasted thousands of competitors that failed. The molecule in that aspirin tablet is the survivor of an astonishing filter.

By the time you finish this chapter, you will be able to look at a headline about a new drug approval, or a drug pulled from shelves, and understand exactly what category of molecular problem just happened. You will know what a receptor is, why the shape of a molecule determines whether it turns something on or shuts it off, and how penicillin, a molecule made by a fungus, accidentally stumbled into one of the most consequential discoveries in medical history. Let's start at the lock.

The receptor model: lock and key

Your body is made of proteins. Not just the proteins in your muscles or your hair, but thousands of kinds of proteins that act as machines, messengers, gates, and sensors. Every cell is studded with proteins that sit in the cell membrane like antennas, waiting for a signal. Those proteins are called receptors, and they are the targets that drugs are designed to hit.

Think of a receptor as a physical object with a very specific shape. Most receptors have a small pocket or groove somewhere on their surface, a cavity just large enough to hold a specific molecule. That molecule, the one the receptor evolved to recognize, is called a ligand. In your body, natural ligands include hormones, neurotransmitters, and other signaling molecules. Adrenaline is a ligand for adrenaline receptors. Serotonin is a ligand for serotonin receptors. When the right ligand slides into the right pocket, the receptor changes shape, and that shape triggers a cascade of events inside the cell.

The lock-and-key model is the oldest way of describing this, and it still works beautifully as an introduction. The receptor is the lock. The ligand is the key. The right key fits, turns, and opens the lock. A wrong key, even one that is almost the right shape, rattles around and fails to turn anything. A drug is an artificial key, one that chemists designed or discovered to fit that specific lock.

The lock-and-key model is a simplification, and we should acknowledge that before going further. In reality, both the receptor and the ligand are flexible. When a ligand approaches a receptor, they both shift slightly to accommodate each other, much like two pieces of clay pressed together that deform to increase their contact. This more accurate picture is called the induced fit. But the core insight remains: shape is everything. A molecule that cannot fit the receptor pocket cannot trigger the receptor. And a molecule that fits more precisely typically triggers a stronger or more specific response.

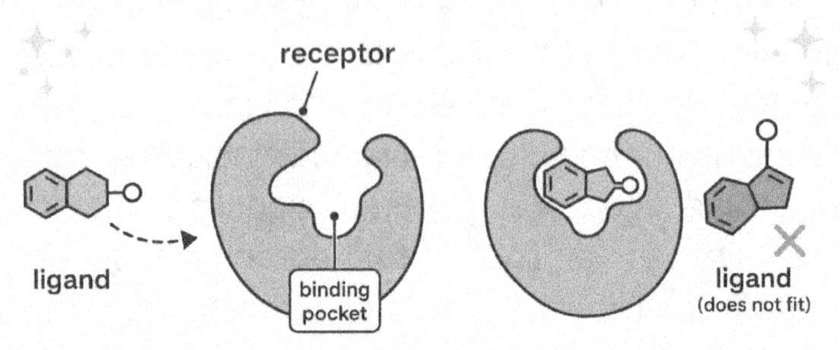

The receptor pocket and the lock-and-key model.

Diagram 6.1: The receptor pocket and the lock-and-key model

The binding pocket is lined with specific amino acids, each one contributing a small attraction to the incoming ligand. Some of those attractions are hydrogen bonds, in which hydrogen atoms on the ligand weakly connect to oxygen or nitrogen atoms on the receptor wall. Some are van der Waals forces, the gentle flickering attractions between any two surfaces that are close enough. Some are hydrophobic effects, the tendency of oily regions to cluster away from water, which pushes certain parts of the ligand snugly against certain parts of the pocket. None of these individual interactions is strong. But together, summed across dozens of contact points, they add up to a binding grip that is firm, specific, and remarkably selective.

This selectivity is the key to drug design and its greatest challenge. You want your drug to fit one receptor pocket precisely. You do not want it fitting thirty other protein pockets in your body, because every off-target binding event is a potential side effect. The goal is specificity: a drug so well-shaped for its target that it ignores everything else. Achieving that is harder than it sounds, because you are designing at the scale of atoms.

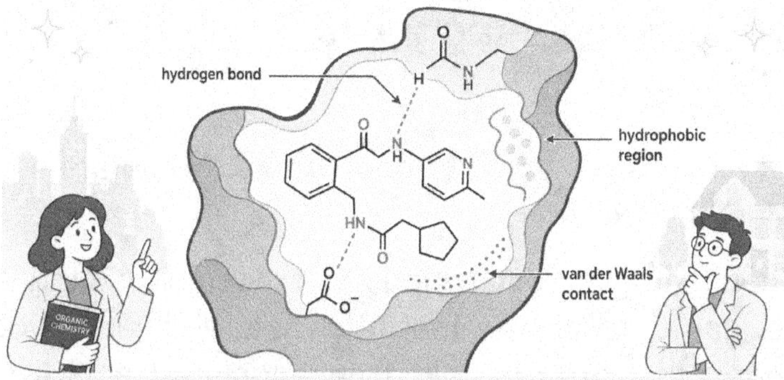

Molecular interactions in the binding pocket.

Diagram 6.2: Molecular interactions in the binding pocket

To make things more interesting, many proteins have more than one binding site. The main binding site, called the orthosteric site, is where the natural ligand binds. But some proteins have secondary sites, called allosteric sites, where a different molecule can bind and indirectly change the receptor's shape, without sitting in the main pocket at all. Allosteric drugs are a growing area of research because they can fine-tune a receptor's behavior rather than simply turning it on or off. That level of control is very attractive when you are trying to treat a condition without disrupting everything else the receptor does.

For now, picture the main picture: a drug molecule floating through your bloodstream, bumping into protein after protein, failing to bind until it finds the one receptor it was built for, and then clicking into place the way a key clicks into a lock. That binding event changes the receptor. And that change is the beginning of the drug's effect.

Agonists vs antagonists: turn on vs block

Not all keys are trying to open doors. Some are trying to jam the lock.

When a drug molecule binds to a receptor and triggers the same response that the natural ligand would trigger, that drug is called an agonist. It acts like a natural messenger but is usually more potent, longer-lasting, or more targeted. Morphine is an agonist at opioid receptors. It binds the same pocket that your body's natural pain-suppressing molecules bind, activates the same machinery, and delivers a much more intense version of the same signal.

When a drug molecule binds to a receptor and blocks it without activating it, that drug is called an antagonist. It sits in the lock and refuses to turn. The natural ligand arrives and cannot get in. Nothing happens. The receptor is silent. The effect of an antagonist is the opposite of that of an agonist: it does not create a new signal; it prevents the natural signal from getting through.

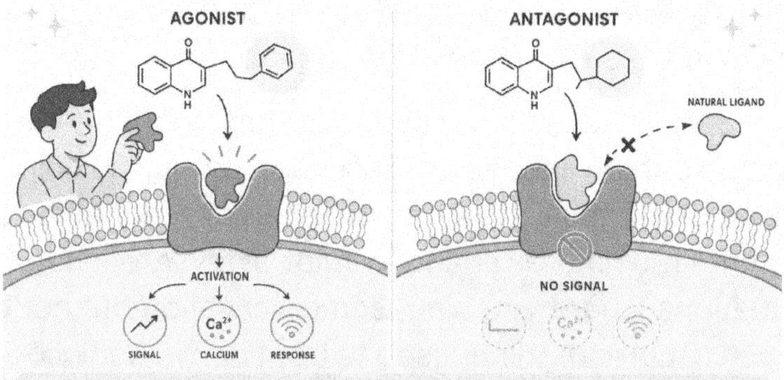

Agonist vs antagonist at a receptor.

Diagram 6.3: Agonist vs antagonist at a receptor

Beta-blockers are among the most widely prescribed antagonists worldwide. They block beta-adrenergic receptors, which normally respond to adrenaline. When adrenaline binds to its receptors without interference, your heart rate climbs, your blood pressure rises, and your body prepares for action. Beta-blockers park themselves in those receptor pockets before adrenaline can get there. The heart beats more slowly and steadily. For someone with high blood pressure or a heart that tends to race, this is exactly what the doctor ordered. Literally.

Antihistamines work the same way, but at a different receptor. Histamine is a molecule your immune cells release when they detect an allergen. It binds histamine receptors in your nasal passages and eyes, triggering swelling, a runny nose, and the general misery of allergy season. Antihistamines are antagonists that block those histamine receptors. With the receptors occupied by the drug, histamine cannot bind to them. The symptoms quiet down. You stop sneezing. You can go outside again.

There is a more nuanced category worth knowing about: partial agonists. These are drugs that bind a receptor and activate it, but only partially, not to the full extent the natural ligand would. They are useful in situations where you want some activation but not too much. Buprenorphine, used to treat opioid addiction, is a partial agonist at opioid receptors. It satisfies the receptor sufficiently to prevent withdrawal but not enough to produce the full effect that creates dependence.

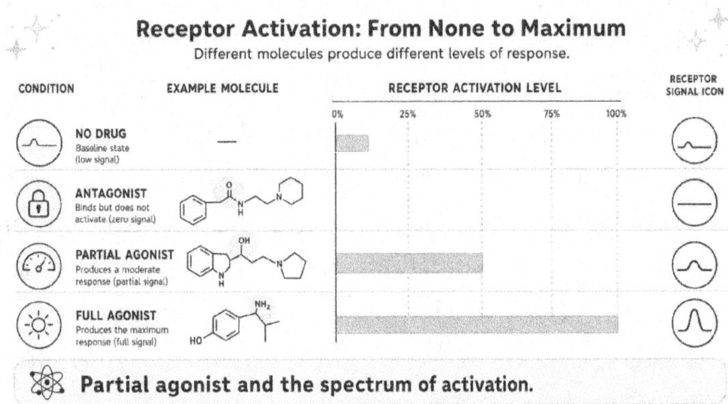

Partial agonist and the spectrum of activation.

Diagram 6.4: Partial agonist and the spectrum of activation

There is another wrinkle. Some drugs are inverse agonists. They do not just block the receptor; they actively push it into a less active state than it would be without any drug. Many receptors exhibit low background activity even in the absence of their natural ligand. An inverse agonist suppresses even that baseline activity. The distinction between an inverse agonist and an antagonist matters clinically, though for a first pass, an antagonist is close enough to get the idea.

The agonist-antagonist framework explains a huge swath of modern pharmacology. When a drug is described as a "blocker," it is almost certainly an antagonist. When a drug is described as a "stimulator" or a "mimic," it is acting as an agonist. Every drug that interacts with a receptor falls somewhere on this spectrum. And where it falls determines almost everything about what it does.

Finding leads: natural products and screening.

Here is the uncomfortable truth about drug discovery in most of history: it was mostly accident, observation, and a great deal of luck dressed up as systematic science afterward.

For most of human history, medicines came from plants. Willow bark tea for fever and pain. Foxglove for heart conditions. Opium poppies for pain relief. Cinchona bark for malaria. Every one of those remedies worked because the plant contained a specific molecule that hit a specific receptor in the human body. The healers who used these remedies did not know about receptors. They knew that the bark worked, and they kept using it.

This approach, which starts with a natural product with a known biological effect and then isolates and studies the active molecule, is called natural product discovery. It is still enormously productive today. Roughly half of all drugs approved in the last fifty years are derived from natural products or were inspired by natural product structures. Nature is a prolific chemist, and natural selection has had billions of years to experiment with molecules that interact with biological systems.

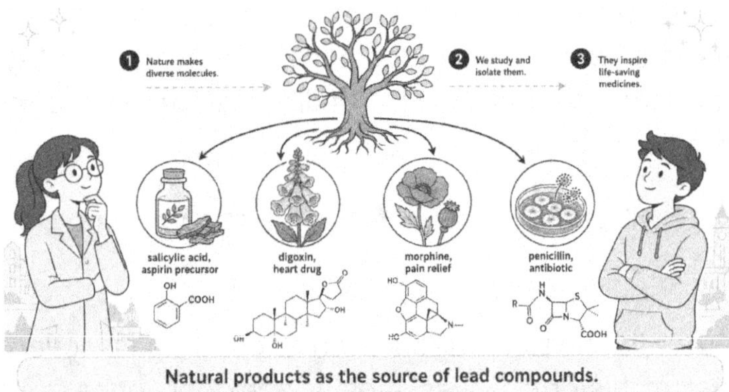

Diagram 6.5: Natural products as the source of lead compounds

The molecule that modern chemists isolate from a natural product is called a lead compound, or just a lead. It is not necessarily the drug itself; it is the starting point. Chemists take the lead molecule and begin modifying it, changing one part at a time, testing each modification to see whether it is more potent, more selective, less toxic, or easier for the body to absorb. This iterative process is called lead optimization, and it can take years.

The second major route to finding leads is high-throughput screening. This is less poetic than discovering drugs in the rainforest, but it is extraordinarily powerful. A pharmaceutical company builds a library of hundreds of thousands of small molecules, synthesized over years of prior research. Then they take a target protein, a receptor they think is involved in a particular disease, and they test every molecule in the library against it. All of them. In parallel. Robotic systems can run tens of thousands of tests per day. When a molecule shows some binding or some activity, it gets flagged for follow-up.

High-throughput screening produces many false positives and many dead ends. But it is systematic. You are not waiting for a lucky observation in a hospital ward or a forest; you are running a structured search. The molecules that survive the initial screening, called hits, are further characterized. The best hits become leads, and leads enter the optimization pipeline.

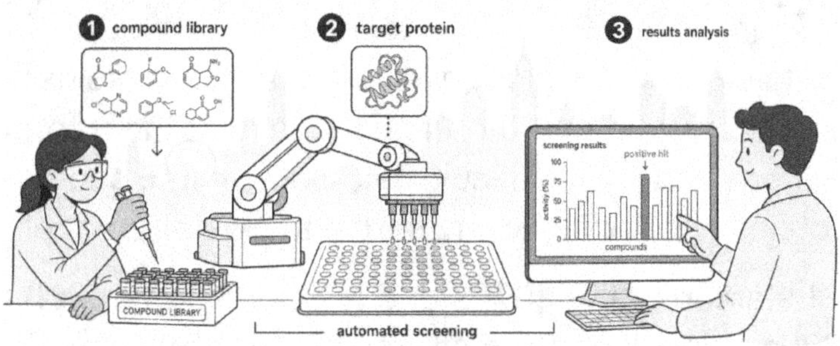

High-throughput screening

Diagram 6.6: High-throughput screening

The third major route, rational design, is the newest and most conceptually satisfying. It gets its own section because it represents a genuine revolution in how we think about making drugs. But before we get there, it is worth pausing to appreciate what high-throughput screening and natural product discovery share: they both start with finding something that works and then asking why. Rational design flips that order. You start with the why and then build the what.

Rational design: the modern era

For most of drug history, the question was: "This molecule works, but why?" Rational design asks the reverse: "We know exactly why this receptor does what it does. Now, what molecule can we build to control it?"

To do rational design, you need a detailed, three-dimensional picture of the target receptor. And for most of the twentieth century, that was simply impossible. Proteins are small, folded into complex shapes, and do not sit still. Getting an accurate picture of a protein's structure requires a technique called X-ray crystallography, which involves coaxing proteins into perfect crystals and then firing X-rays at them to map their positions. It is difficult, time-consuming, and not all proteins cooperate. Many refuse to crystallize at all.

But starting in the 1970s and accelerating through the 1980s and 1990s, structural biology gave us increasingly detailed pictures of receptor binding pockets. And once chemists could see the pocket, they could design keys to fit it. This is the beginning of modern rational drug design.

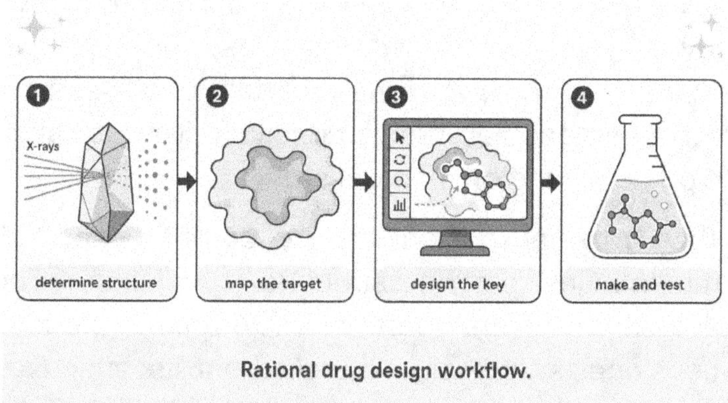

Diagram 6.7: Rational drug design workflow

The process goes roughly like this. Structural biologists solve the crystal structure of the target protein. Computational chemists examine the binding pocket, identify all of its chemical features (where are the hydrogen bond donors? where are the hydrophobic regions? what is the volume and shape of the cavity?), and then use software to search for or generate molecules that should fit. Those candidate molecules are synthesized in the lab and tested. The ones that bind go into optimization.

One of the most celebrated early successes of rational drug design was the development of HIV protease inhibitors in the 1990s. HIV is a virus that needs to cleave certain large proteins into smaller functional pieces to replicate. The enzyme that carries out the cleavage, called HIV protease, has a binding pocket where the protein substrate sits just before being cut.

Researchers solved the structure of HIV protease, mapped the pocket, and designed molecules specifically shaped to fit in the pocket and prevent cleavage. The result was a class of drugs that transformed HIV from a near-certain death sentence into a manageable chronic condition. That is rational design delivering on its promise.

Today, rational design has a powerful new ally: computational protein structure prediction. Tools developed in recent years can now predict the three-dimensional shape of a protein from its amino acid sequence with remarkable accuracy. This feat took years of experimental crystallography to achieve. This has dramatically expanded the universe of targets that can be addressed by rational design, because you no longer need a crystal structure to start designing. You can predict the structure and work from there.

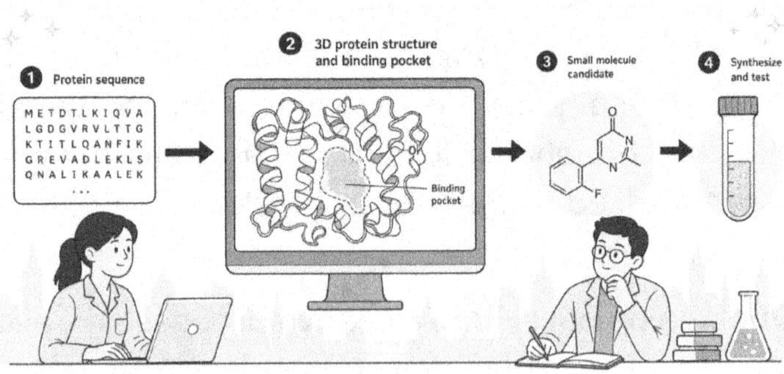

Computational drug design.

Diagram 6.8: Computational drug design

The frontier of rational design now includes designing drugs for targets that were previously considered "undruggable": proteins whose binding sites are too flat, too flexible, or too large for a conventional small molecule to bind. New strategies, including molecules that work at allosteric sites, molecules that degrade the target protein entirely rather than blocking it, and molecules that recruit the body's own cellular garbage-disposal machinery, are expanding what rational design can achieve. The field is moving faster now than at any point in its history.

The drug pipeline: lab bench to medicine cabinet

Finding a molecule that binds a target is the beginning of the story. Getting it into a patient's medicine cabinet is a decade-long obstacle course. The pharmaceutical development pipeline is one of the most rigorous and expensive filters ever devised.

The pipeline has several formal stages, each designed to answer a specific question. And at every stage, most candidates fail. The numbers are sobering. Of roughly ten thousand molecules that enter the early stages of investigation, perhaps one will eventually reach approval as a drug. The rest are eliminated for being too toxic, insufficiently effective in animals or humans, impossible to formulate as a stable pill, cleared from the body too quickly, or for any of a hundred other reasons.

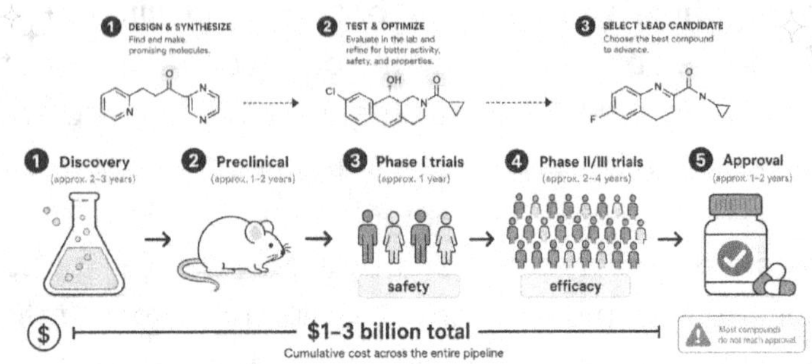

The drug development pipeline.

Diagram 6.9: The drug development pipeline

The discovery phase is where we have spent most of this chapter. You identify a target, find lead compounds through natural products, screening, or rational design, and optimize the best leads.

Preclinical testing is where the optimized lead compound is tested for the first time in living systems, but not humans. This involves cell cultures (testing whether the drug works in isolated cells growing in a dish) and animal models (usually rodents, sometimes larger animals). Preclinical testing seeks to answer one question: Does it work in a living organism? And is it dangerously toxic? Both questions are necessary. A drug that works beautifully but destroys the liver fails here.

Phase I clinical trials are the first time a new drug is given to human beings. The participants are usually a small number of healthy volunteers, not patients with the disease. Phase I is almost entirely about safety. What happens when a person takes this compound? What doses can be tolerated? How quickly does the body absorb and excrete it? At what dose do side effects appear? Phase I is not trying to prove the drug works; it is trying to prove it does not immediately harm people.

Phase II trials enroll a larger group of patients who actually have the disease or condition the drug is meant to treat. Now you are asking: Does it work? Phase II begins to gather preliminary evidence of efficacy and continues refining the safety profile.

Phase III trials are the largest and most expensive stage. Hundreds or thousands of patients, often at multiple sites across multiple countries, receive either the drug or a placebo (or the current standard of care). Phase III is designed to give definitive statistical evidence that the drug works and that its benefits outweigh its risks. These trials can last several years. They cost enormous sums. And they are still where many drugs fail, either because efficacy is not as strong as Phase II suggested, or because rare side effects emerge only after treating enough patients.

If a drug survives Phase III, the company submits a mountain of data to regulatory agencies, such as the U.S. Food and Drug Administration. The review process itself can take another year or more. Approval is not guaranteed even at this stage; regulators can request additional studies or reject the application if they find the benefit-risk balance inadequate.

After approval, surveillance continues. Phase IV, sometimes called post-marketing surveillance, tracks the drug as millions of patients use it in conditions that clinical trials never perfectly replicated. This is how rare but serious side effects are sometimes discovered years after a drug is on the shelves. It is also why drugs occasionally get withdrawn even after approval. The filter never fully closes.

All of this is why drug development is expensive. Every one of those ten thousand early candidates costs money to synthesize, test, and characterize. The clinical trials for each one that reaches that stage cost more. The infrastructure, the regulatory expertise, the manufacturing development, all of it compounds. The billion-dollar price tag is not the result of greed; it is the result of an extraordinarily high-stakes process of elimination.

Origin stories: aspirin, penicillin, ibuprofen

The abstract machinery of drug development becomes a lot more vivid when you see it in specific cases. Three drugs in particular illustrate the range of ways drugs get discovered: one was an accident, one was a lucky observation, and one was a deliberate corporate search.

Aspirin: from willow to pill

People have chewed willow bark for pain relief for thousands of years. Ancient Egyptians, ancient Greeks, and indigenous peoples across multiple continents independently discovered that it worked. What they did not know was why.

In the early nineteenth century, chemists began isolating the active ingredient. In 1828, a French pharmacist named Henri Leroux isolated a compound called salicin from willow bark. Salicin breaks down in the body to release salicylic acid, the molecule actually responsible for the pain relief and fever reduction. Salicylic acid worked, but it was harsh: it caused severe stomach irritation and had an unpleasant taste.

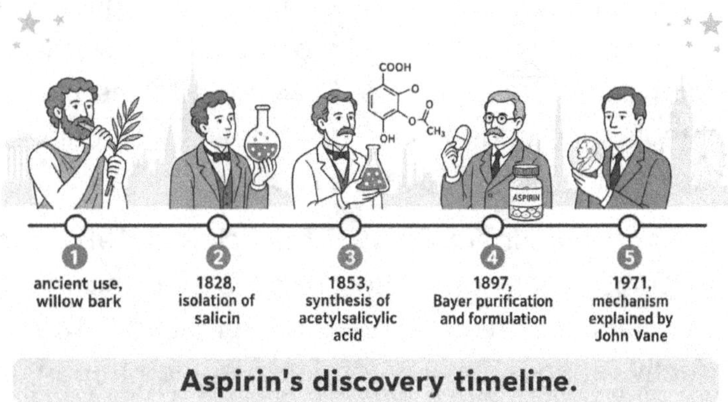

Diagram 6.10: Aspirin's discovery timeline

In 1853, the French chemist Charles Gerhardt synthesized acetylsalicylic acid, the molecule we now call aspirin, by adding an acetyl group to salicylic acid. This modification made the compound more stable and less irritating to the stomach lining. But Gerhardt did not pursue it commercially; he moved on to other projects. It took another half-century before Felix Hoffmann, a chemist at the German company Bayer, re-synthesized aspirin in a pure form in 1897. It convinced his employer to develop it as a product.

Here is the remarkable thing: aspirin went from a folk remedy to a commercial drug to a global bestseller without anyone understanding how it worked. The mechanism of action was not explained until 1971, when British pharmacologist John Vane showed that aspirin inhibits a family of enzymes called cyclooxygenases (COX). COX enzymes produce prostaglandins, which are signaling molecules that amplify pain, promote inflammation, and trigger fever. Aspirin binds to the active site of the COX enzyme and permanently blocks it. No COX activity means less prostaglandin production, which means less pain, less inflammation, and reduced fever.

Vane won the Nobel Prize in Medicine in 1982, more than eighty years after the drug reached the market. The molecule arrived before the explanation. That is a perfectly reasonable way for discovery to happen, and it happened with aspirin for over a century.

Ibuprofen, the other common pain reliever in your medicine cabinet, also inhibits COX enzymes. But where aspirin permanently turns off the enzyme, ibuprofen binds reversibly, settling into the active site and blocking it without forming a permanent bond. When ibuprofen clears from the body, the COX enzyme recovers. This is one reason ibuprofen has a shorter duration of action and why taking multiple doses throughout the day is necessary. It is also why ibuprofen and aspirin, despite targeting the same pathway, are not identical in their effects or side-effect profiles.

Penicillin: the accidental antibiotic

Alexander Fleming was not a tidy man. In 1928, he returned from a vacation to find that one of his bacterial culture plates had been contaminated by mold. The mold, a species called Penicillium notatum, had created a clear zone around itself where bacteria refused to grow. Fleming, trained as a bacteriologist, recognized immediately that the mold must be producing something that killed bacteria. He isolated the mold, studied the compound it produced, and called it penicillin.

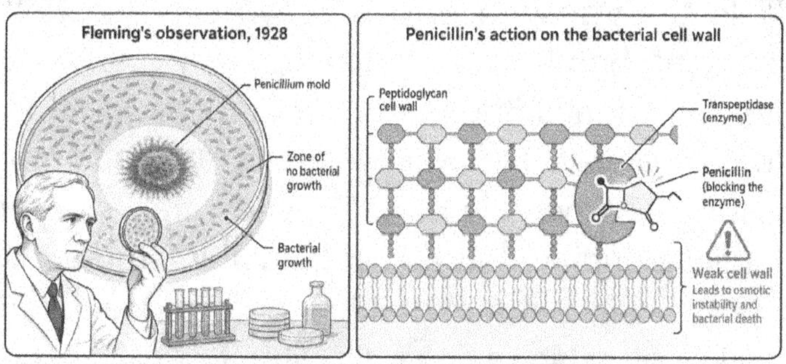

Penicillin's mechanism and discovery.

Diagram 6.11: Penicillin's mechanism and discovery

Fleming published his observation but lacked the resources and expertise to purify and further develop penicillin. The compound was unstable, difficult to isolate, and, at that point, a curiosity rather than a treatment. It took another decade, and the urgency of World War II, before Howard Florey and Ernst Boris Chain at Oxford University developed methods to purify penicillin in meaningful quantities. The first clinical trials in the early 1940s were stunning. Patients with severe bacterial infections that would have killed them recovered rapidly. Penicillin was mass-produced in the United States for Allied troops, and its impact on wartime mortality was enormous.

What makes penicillin elegant from a chemistry standpoint is its mechanism. Bacteria need to build and maintain their cell walls, a rigid mesh of cross-linked molecules that holds the bacterial cell together under its own internal pressure. A specific enzyme, called transpeptidase (sometimes called penicillin-binding protein), stitches those cross-links together. Penicillin fits into the transpeptidase's active site and forms a permanent bond, inactivating the enzyme. Without active transpeptidase, the bacterium cannot repair or build its cell wall. The wall weakens, the cell's internal pressure pushes outward, and the bacterium essentially bursts. It is a spectacularly targeted killing machine: it hits a target that bacteria have and human cells do not, which is exactly why it can be lethal to bacteria and safe for us.

The discovery of penicillin launched the entire era of antibiotics. Within two decades, chemists had identified dozens of related compounds from other mold species and modified them systematically to create broader-spectrum antibiotics. These drugs killed a wider range of bacterial species. The natural product was the lead; the medicinal chemists did the optimization. The logic is identical to the process described earlier in this chapter.

Ibuprofen: the corporate search

If aspirin was accidental and penicillin was a lucky observation, ibuprofen was the result of a deliberate search. In the late 1950s, a British pharmaceutical company, Boots Company (then known for pharmaceuticals), sought a better anti-inflammatory drug. Aspirin worked but caused stomach problems. Corticosteroids were more powerful but had serious side effects with long-term use. The company believed there was room for a safer, better-tolerated alternative.

A team led by chemist Stewart Adams and his colleagues began systematically testing a series of compounds related to phenylacetic acid, a simple aromatic molecule. They were looking for anti-inflammatory activity with an acceptable safety profile. The screening was methodical: synthesize a candidate, test it in an animal model of inflammation, measure the result, modify the structure, repeat. This is lead optimization in practice.

After testing hundreds of compounds over several years, the team identified ibuprofen, which chemists at the time gave the decidedly unromantic name 2-(4-isobutylphenyl)propionic acid (a long string that describes the molecule's backbone and the side groups attached to it, without conveying any of its future fame). Adams reportedly first tested it on himself to check its safety, taking a dose when he had a hangover-induced headache. It worked. Ibuprofen received approval in the United Kingdom in 1969, first as a prescription drug, and was approved as an over-the-counter medication in the United States in 1984. It is now one of the most widely used drugs on Earth.

The ibuprofen story is notable because it shows systematic drug discovery working as intended: a clear medical need, a defined class of targets, a structured search, and a molecule that emerged from careful chemistry. No lucky contamination. No ancient folk remedy. Just the method.

The future: personalized medicine

Every drug discussed in this chapter is a blunt instrument by the standards of where medicine is heading. Aspirin inhibits all COX enzymes in every tissue of your body, which is why it reduces fever, reduces inflammation, prevents blood clotting, and irritates your stomach lining, all at once. It does not know where you need it. It cannot distinguish between the COX enzymes in your joints and the ones in your stomach wall. This is the core limitation of most traditional drugs: they are chemically precise but biologically non-specific because the target they hit is present in many tissues, not just the ones causing you trouble.

The future of medicine is moving toward far greater specificity. Not just receptor-specific drugs, but patient-specific drugs. The vision is a world where the drug you receive is designed not just for your disease but for your particular molecular version of that disease.

Cancer medicine is leading this shift. Specific mutations in specific genes drive many cancers. Two patients with the same type of cancer might have tumors driven by completely different molecular defects. A drug that targets a single driver mutation will be effective for the patient with that mutation and ineffective for the patient whose tumor is driven by a different mutation. Precision oncology, which matches targeted drugs to specific tumor genetics, is already extending and saving lives. Drugs like imatinib, which target a specific abnormal protein produced by a chromosomal defect found in chronic myelogenous leukemia, achieve remission rates that were unimaginable with older chemotherapy approaches.

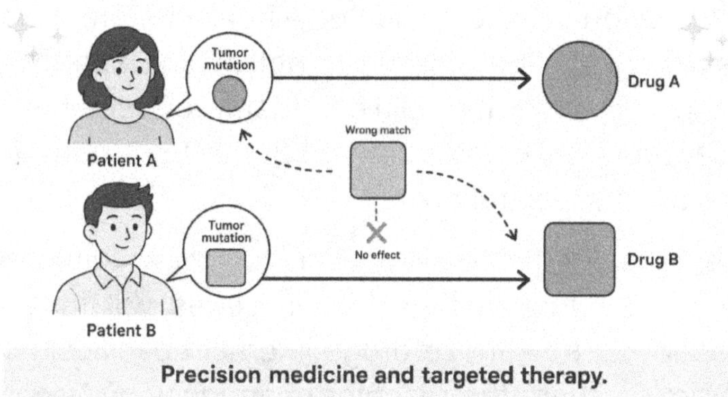

Diagram 6.12: Precision medicine and targeted therapy

Beyond cancer, personalized medicine intersects with pharmacogenomics, the study of how your genetic makeup affects how your body processes drugs. Some people metabolize certain drugs very quickly because they carry extra copies of the gene for a drug-processing enzyme. For them, a standard dose clears the body before it has time to work. Others are slow metabolizers; a standard dose builds up to toxic levels. Knowing a patient's metabolizer status before prescribing can mean the difference between an effective treatment and a dangerous one.

The convergence of structural biology, computational chemistry, genomics, and gene sequencing technology is compressing the timeline of drug discovery in ways that were not possible even a decade ago. New drug targets that would have taken years to characterize can now be mapped in months. Candidate molecules that would have taken enormous chemical effort to synthesize are being designed computationally and evaluated virtually before a single flask is used. The pipeline is not getting shorter yet, mostly because the clinical trial requirements are necessarily slow and human, but the front end of the pipeline is accelerating.

One note for those who want to follow this thread further: a future volume in this series will take you deeper into how these drugs interact specifically with the brain, where the receptor pharmacology becomes even more intricate, and the stakes become even more personal.

Every pill in your medicine cabinet is the survivor of a molecular obstacle course that took years, cost an extraordinary amount of money, and required the combined effort of chemists, biologists, pharmacologists, physicians, regulators, and statisticians. The next time you shake two aspirin from a bottle and swallow them with a glass of water, you are using the distilled product of about 150 years of organic chemistry. Those molecules know where they are going.

Try this

Next time you look at a drug label, read the "mechanism of action" section (it is usually in the package insert, if not on the bottle itself). See if you can identify whether the drug is an agonist or an antagonist. If it "blocks" something, it is an antagonist. If it "stimulates" or "activates" something, it is an agonist. You now have the vocabulary to understand what that sentence means at the molecular level.

Takeaways

• You can explain how a drug finds its molecular target: the drug is a shaped molecule that fits a specific pocket on a receptor protein, the way a key fits a lock.

• You can distinguish an agonist from an antagonist: an agonist binds a receptor and activates it; an antagonist binds the same pocket and silences it.

• You can describe three routes to finding a drug candidate: natural product discovery (starting with something that works in nature), high-throughput screening (testing enormous libraries of molecules against a target), and rational design (starting from the receptor structure and designing a molecule to fit it).

• You can explain why the drug development pipeline takes a decade and costs over a billion dollars: thousands of candidates fail at each stage of a multiphase filter designed to prove both safety and efficacy in human beings.

• You can tell the origin stories of three landmark drugs: aspirin emerged from willow bark folk medicine and took 150 years to have its mechanism explained; penicillin was an accidental observation by Fleming that wartime urgency turned into a revolution in medicine; ibuprofen was the product of systematic corporate pharmaceutical research.

• You can articulate the central promise of personalized medicine: matching a drug not just to a disease but to the specific molecular defect driving a particular patient's version of that disease.

• You can read a headline about a drug approval or recall and recognize the molecular category of what happened: was the drug failing to bind its target, binding off-target proteins, or proving unsafe in a broader population?

The molecule that is right for you might still be waiting to be built. And the people who will build it are working with the same tools this chapter has described, the same logic of shape, fit, target, and key. The next chapter takes that logic of molecular design and applies it to a different kind of molecule entirely: the polymer. Where drugs are built to be small and precise, polymers are built to be enormous by repetition. Where a drug molecule might have thirty atoms, a polymer chain might have thirty thousand. The same carbon-based chemistry, on a completely different scale. Welcome to the world of plastics.

7 Miracle and Menace: Plastics and Polymers

Look around the room you are in right now.

If there is a carpet underfoot, it is likely polyester. The bag hanging off your chair is probably nylon. The bin in the corner is polyethylene. The yogurt cup on the desk is polypropylene. There is a PVC pipe hiding inside the walls. The takeout box sitting on the counter is polystyrene. Every single one of those materials has a different name, texture, purpose, and country of manufacture. And every single one of them is the same trick.

Take a small molecule. Repeat it: ten thousand times, a hundred thousand times, sometimes a million times. Get something completely new.

That trick has a name. It is called polymerization, and the product is a polymer. It is not an exaggeration to say that this one idea reshaped the twentieth century more than almost anything else. Polymers are why modern surgery exists. They are why we can send satellites into orbit, store food for months, and dress ourselves for almost nothing. They are also why there is plastic in the deepest ocean trench, in the air over the Himalayas, and almost certainly in your bloodstream right now.

Understanding polymers means understanding both the miracle and the menace. By the end of this chapter, you will be able to explain what a polymer is, why plastics are so chemically stubborn that they persist for centuries, and what the molecular difference is between a plastic bag that will outlive your grandchildren and a bioplastic that breaks down in a compost bin. You will also understand why your body is, in a very real sense, a walking polymer factory. DNA is a polymer. Proteins are polymers. Cellulose is a polymer. Life itself discovered this trick long before Leo Baekeland made the first synthetic plastic in 1907.

Let us start at the beginning, with one molecule and one bond.

What a polymer really is

A polymer is a long-chain molecule made by linking smaller molecules together, end to end, over and over again. The small molecules are called monomers, from the Greek for "one part." The polymer is the chain they form when you connect them. That is it. That is the whole idea.

Here is the best analogy for it: think about LEGO bricks. A single LEGO brick is not very impressive. It sits in your palm and does nothing in particular. But give a child ten thousand identical bricks and enough time, and you get something entirely new, something with structure, with properties, with presence. The same is true of monomers. One molecule of ethylene is a gas. It floats away. You cannot feel it, cannot build anything with it, cannot wrap a sandwich in it. But link ten thousand ethylene molecules together, and you get polyethylene, a tough, waxy, endlessly versatile solid.

The LEGO brick gained a superpower just by finding its neighbors.

Ethylene is the simplest possible monomer. It is a molecule with two carbon atoms joined by a double bond, each bearing two hydrogen atoms. That double bond is the key. During polymerization, the double bond opens up. The two carbons, which were sharing four electrons in their double bond, now each reach out and grab onto the next ethylene molecule. They form a new single bond. The process repeats down the line. One by one, each new ethylene finds its place in the growing chain, and the double bond disappears, replaced by a long backbone of carbon-carbon single bonds. The finished chain is polyethylene.

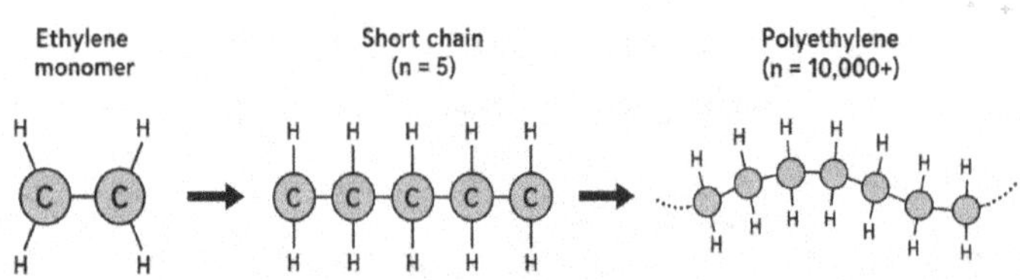

From ethylene monomer to polyethylene chain.

Diagram 7.1: From ethylene monomer to polyethylene chain

The number of times the monomer repeats is called the degree of polymerization, and it matters enormously. Short chains of ethylene, a few dozen units long, give you a liquid; chains a few hundred units long give you a soft wax. Chains in the tens of thousands give you the firm, translucent material you recognize as a shopping bag. Chains in the hundreds of thousands give you the hard, white material used in pipes and cutting boards. Same molecule. Same bond. Just more of it. The length of the chain is a dial that chemists can turn to get almost any mechanical property they want.

This is the central insight of polymer chemistry: you are not changing the atoms, you are changing the architecture. Carbon chemistry is endlessly creative, not because carbon does different things in different contexts, but because the same things done at different scales, and in different arrangements, produce wildly different results.

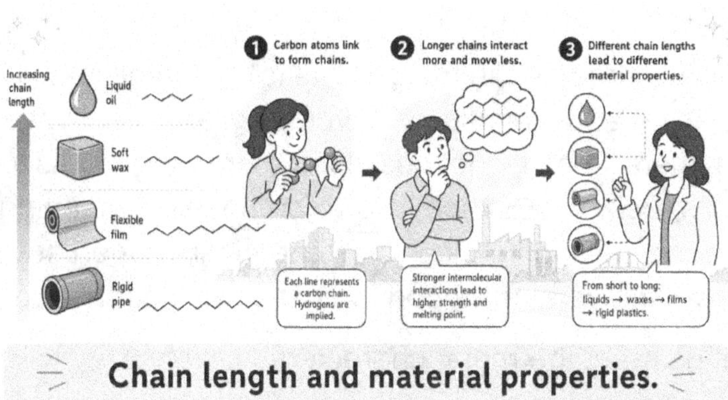

Diagram 7.2: Chain length and material properties

The process by which ethylene monomers link together is called addition polymerization. The double bond opens and adds to the next unit. No atoms are lost. No byproducts are formed. You start with n monomers and end with one chain that contains all n units. Tidy.

Efficient. This is why addition polymers can be made at an industrial scale with very little waste from the reaction itself.

There is a second type, called condensation polymerization, which works differently. In condensation polymerization, each time two monomers join, they also expel a small molecule, usually water. This is how nylon is made, how polyester is made, how your body makes proteins. We will get back to those. For now, the point is that "polymer" is not a single recipe but a family of ideas united by a single theme: the chain.

The major plastics and their personalities

Plastics are not one thing. "Plastic" is a family name for a large and genuinely diverse group of synthetic polymers. Each one has its own monomer, its own chain architecture, and its own personality, meaning its own set of mechanical and chemical properties that determine what it is good for and what it cannot do.

Let us meet the major characters.

Polyethylene (PE) is the most produced plastic in the world. Its monomer, ethylene, comes primarily from natural gas or crude oil. Polyethylene is waxy, flexible, and chemically inert. It does not react with water, acids, or most solvents. This makes it perfect for bags, bottles, food wrap, and containers. There are actually several grades of polyethylene, depending on how the chains are arranged. Low-density polyethylene (LDPE) has many branches off the main chain, which prevent the chains from packing tightly together. This gives you a soft, flexible film, the kind used in sandwich bags. High-density polyethylene (HDPE) has straighter chains that pack more tightly, giving you a stiffer, denser material used in milk jugs, detergent bottles, and cutting boards.

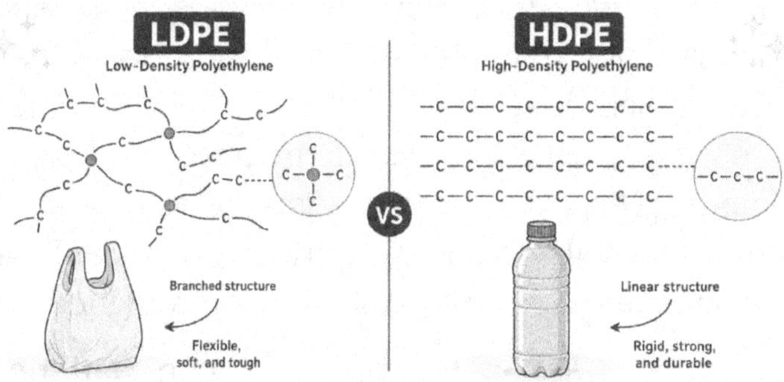

LDPE vs HDPE chain architecture.

Diagram 7.3: LDPE vs HDPE chain architecture

Polypropylene (PP) is ethylene's close cousin. The monomer is propylene, which is like ethylene but with one extra methyl group sticking off one of the carbons. That one extra carbon changes everything. Polypropylene is harder and more heat-resistant than polyethylene. It can withstand boiling water without deforming, which is why it is used in yogurt cups, dishwasher-safe food containers, and the living-hinge mechanisms on flip-top bottle caps. The molecular reason is that the methyl groups on every other carbon of the backbone create steric bulk, which physically blocks and locks the chain into a stiffer conformation.

PVC, polyvinyl chloride, is the one with chlorine. Its monomer, vinyl chloride, is an ethylene molecule with one hydrogen replaced by a chlorine atom. That chlorine is heavy and electronegative. It makes the polymer stiffer and denser than polyethylene. Rigid PVC is used for pipes, window frames, and credit cards. When you add plasticizers (small molecules that work their way between the chains and act as lubricants), you get soft PVC: shower curtains, garden hoses, artificial leather. PVC is chemically durable and fire-resistant, which is why it replaced lead pipes in buildings. The irony is that some of the plasticizers used to make flexible PVC, particularly a class called phthalates, have raised health concerns of their own.

Polystyrene (PS) is the one that can puff up. Its monomer is styrene, an ethylene derivative with a benzene ring attached. The polymer chains are stiff because the large benzene rings on every other carbon atom resist rotation. Solid polystyrene is the clear plastic used in disposable cups and CD jewel cases. Expanded polystyrene, what most people call Styrofoam, is the same polymer with tiny air bubbles trapped inside, which makes it an excellent insulating and cushioning material. It is very light, which is why it is beloved for shipping electronics and despised by marine biologists.

PET (polyethylene terephthalate) is the plastic used for beverage bottles. This one is a condensation polymer, made by linking two different monomers that, when joined, expel water at each junction. PET is clear, strong, and has good gas-barrier properties, meaning carbon dioxide does not escape quickly through the walls of a PET bottle. Those same properties that make it ideal for soda bottles also mean it lingers in the environment for a very long time.

Nylon deserves its own paragraph because it was, in many ways, the polymer that proved the concept. Invented at DuPont in 1935 by Wallace Carothers, nylon was the first truly synthetic fiber. Before nylon, fibers came from nature: cotton, wool, silk. Carothers showed that you could build a polymer that had all the strength and flexibility of silk without a single silkworm involved. Nylon is a polyamide, meaning its chains are linked by amide bonds, the same kind of bond that links amino acids together in your proteins. That is not a coincidence. Carothers understood protein chemistry and deliberately mimicked it.

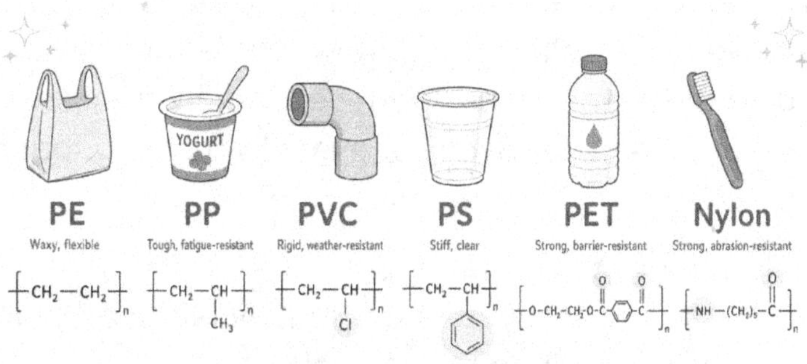

The polymer personality lineup.

Diagram 7.4: The polymer personality lineup

The single most important thing to understand about all of these plastics is that carbon-carbon bonds dominate their backbones. That fact, unremarkable as it sounds, is the molecular explanation for almost everything that follows.

Nature's own polymers: DNA, proteins, cellulose

Before we invented plastics, life had been making polymers for about four billion years. And life's polymers are, in their own way, more impressive than anything we have managed to synthesize.

Let us start with the most famous one.

DNA is a polymer. This is not a metaphor or a simplification. It is literally true. DNA is a long chain molecule made by linking together monomers called nucleotides, end to end, into a chain that can be billions of units long. There are four nucleotides: adenine, thymine, guanine, and cytosine. That is the entire alphabet. Four letters. And from four letters, arranged in different sequences, you get the instructions for every protein in every cell of every living thing that has ever existed.

Think about what that means for a moment. The difference between you and a fruit fly is not a different set of chemical elements. It is not even a different set of polymer building blocks. It is a different sequence of the same four monomers along a polymer chain. The genetic code is a language written in a four-letter alphabet on a polymer backbone, and the message is several billion letters long. For sheer information density and functional elegance, DNA is the most impressive polymer ever made.

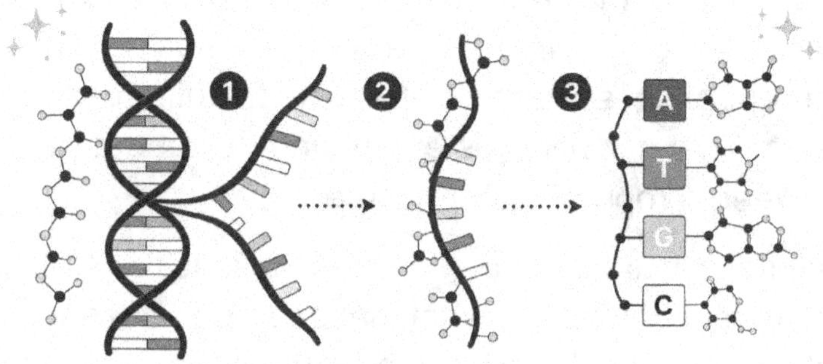

DNA as a four-letter polymer.

Diagram 7.5: DNA as a four-letter polymer

Proteins are also polymers. Their monomers are amino acids, and there are twenty of them in the standard biological toolkit. Unlike DNA, where the four bases are quite similar in structure, the twenty amino acids have very different side chains: some are acidic, some basic, some oily, some water-loving. This variation is what gives proteins their functional diversity. Link glycine, alanine, and valine together one way, and you get a structural fiber like collagen. Rearrange the sequence and vary the amino acids, and you get hemoglobin, which carries oxygen in your blood. Change it again, and you get insulin, which regulates blood sugar. Change it again, and you get amylase, which digests starch in your saliva right now.

The bond that links amino acids together is the amide bond, the same one found in nylon. This is why nylon is sometimes called a synthetic protein: it mimics, in simplified form, the backbone chemistry that life uses to build its own structural and functional polymers. The difference is that a nylon chain has only one type of repeating unit, whereas a protein has a specific, nonrepeating sequence of up to 20 different amino acids. That specificity is everything. It is the difference between a rope and an enzyme.

Cellulose rounds out the trio. Cellulose is the structural polymer of plants. It is what makes wood hard and plant cell walls rigid. Its monomer is glucose, an ordinary sugar. Glucose is the same molecule your cells burn for energy. But when glucose monomers link together in the particular orientation found in cellulose, they form a chain that is extraordinarily rigid and strong, and that your digestive system cannot break down. That is why you cannot digest wood. The glucose is all there, but the bonds holding it together are arranged in a way that human gut enzymes cannot attack.

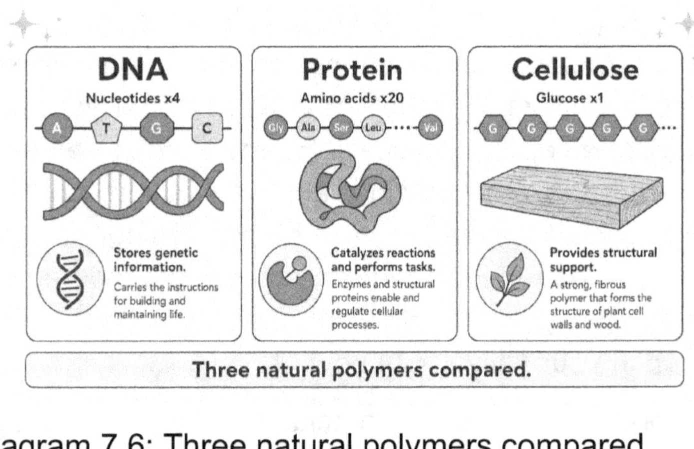

Diagram 7.6: Three natural polymers compared

There is a different cellulose-linked polymer called starch, made from the same glucose monomer but with a different bond orientation. Starch can be digested. The distinction between starch and cellulose is a single geometric difference in how the glucose units join. That tiny molecular difference is the reason a potato is food and a piece of wood is not.

The lesson from nature's polymers is not just that the concept of polymers is universal. It is that sequence that matters. Life uses the polymer architecture to store and execute information. One sequence of monomers is a structural fiber. A different sequence is a catalyst. Yet another sequence is a hormone. The possibilities are effectively limitless. When we make synthetic polymers, we mostly use simple repeating sequences. Life uses the same backbone chemistry to write novels.

Why don't plastics biodegrade

You have probably heard that a plastic bag takes 500 years to decompose. The number varies depending on who you ask, and it is difficult to measure with precision since synthetic plastics have not been around for 500 years, so nobody has watched it happen. But the general principle is correct, and the reason is written in the molecular structure of the bag itself.

Biodegradation occurs when microorganisms, primarily bacteria and fungi, break down a material by attacking its chemical bonds with enzymes. Enzymes are molecular machines, each shaped to recognize and cleave a specific type of chemical bond. Biological systems have evolved enzymes that can cleave ester, amide, and glycosidic bonds (those in cellulose and starch), among others. What they have not evolved enzymes for is the formation of long chains of carbon-carbon single bonds.

This is the molecular explanation for plastic persistence. Polyethylene, polypropylene, and polystyrene all have backbones made of carbon-carbon single bonds, with hydrogen atoms attached. Carbon-carbon single bonds are strong. They do not react with water. They do not react with oxygen at room temperature. They do not react with the acidic environment of a landfill. There is nothing in the molecular structure for an enzyme to grab onto. No ester. No amide. No glycosidic linkage. Just carbons, holding hands with carbons, attached to hydrogens.

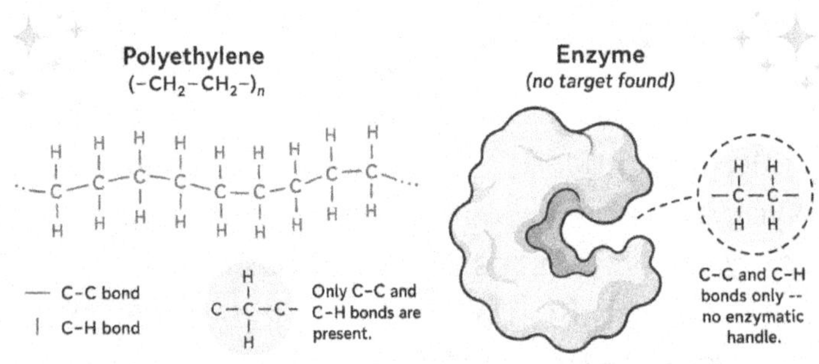

Why enzymes cannot attack polyethylene.

Diagram 7.7: Why enzymes cannot attack polyethylene

Think of it this way. Imagine you are trying to open a door, and every door in the building has a handle except this one. It is completely flat. You can push and pull, but there is nothing to grip. That is what a bacterium faces when it encounters a polyethylene chain. The molecular tools it has evolved to break down organic matter do not fit.

Now, polyethylene does eventually break down. Ultraviolet light from the sun slowly attacks C-H bonds, causing the chain to fragment over decades. Mechanical stress breaks chains apart, too. But this is physical fragmentation, not chemical breakdown. The polymer is not being metabolized into carbon dioxide and water. It is being broken into smaller and smaller pieces of the same plastic, which brings us to the next problem.

PET and nylon do have bonds that enzymes can potentially attack. PET contains ester bonds. Nylon contains amide bonds. Some microorganisms have evolved, quite recently in evolutionary terms, enzymes that can attack these bonds. In 2016, scientists discovered a bacterium called Ideonella sakaiensis that can break down PET using an enzyme now dubbed PETase. This was a significant discovery, and it has sparked a wave of engineering research for faster, more effective plastic-eating enzymes. But even PETase works slowly and only on PET. The all-carbon backbone polymers remain stubbornly immune.

The persistence of plastics is not an accident or a design flaw in the usual sense. It is a direct consequence of the chemical stability that makes plastics so useful in the first place. The same molecular inertness that means a polyethylene pipe does not corrode, a PVC window frame does not rot, and a polystyrene cup does not dissolve in your coffee is the molecular inertness that means these materials accumulate in the environment. Stability is the feature. Persistence is the cost.

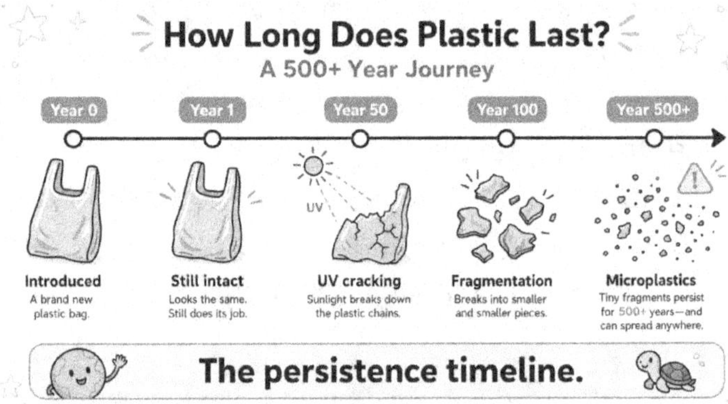

Diagram 7.8: The persistence timeline

Microplastics: how they form and where they end up

If you had to identify the moment when the public conversation about plastic shifted from cautious concern to genuine alarm, it might be the discovery of microplastics. Microplastics are plastic particles smaller than five millimeters. Some of them are far smaller than that, in the nanometer range, which puts them in the same size category as a virus.

They form through exactly the process described above. Physical forces, such as sunlight, wave action, abrasion, and freeze-thaw cycles, break large plastic items into fragments. Each fragment retains the chemical identity of the original polymer. It is still polyethylene, polypropylene, or PET. It has just been reduced in size. This process can continue essentially without limit because there is no chemical reaction to stop it. The chain's chemistry is unchanged by fragmentation. You cannot grind polyethylene fine enough for it to stop being polyethylene.

Microplastics also enter the environment directly. Microbeads, which were used in exfoliating facial scrubs and toothpastes for decades, are manufactured from microplastics. Synthetic fibers shed from clothing in the washing machine are microplastics. Tire wear particles are microplastics. Car tires are a significant fraction of global microplastic emissions, a fact that rarely comes up in discussions that focus exclusively on plastic bottles.

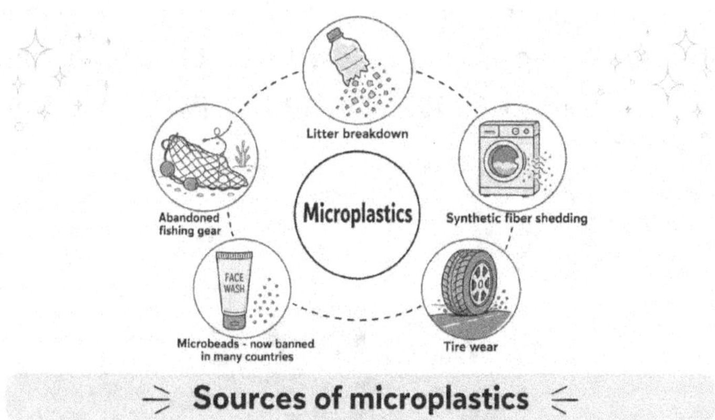

Diagram 7.9: Sources of microplastics

Where do microplastics end up? The short answer is: everywhere. Microplastics have been found in the Mariana Trench, the deepest point in the ocean, at a depth of nearly eleven kilometers. They have been found in Arctic sea ice. They have been found on the summit of Mont Blanc. They have been found in the placentas of unborn babies, in human lung tissue, and in blood samples. Detection methods have improved rapidly in recent years, and with each advance, researchers find microplastics in places they did not expect.

What they do there is an active area of research, and it is worth being precise about what is known and what is uncertain. The physical presence of particles in tissues can cause inflammatory responses. Smaller particles can cross cell membranes in ways that larger particles cannot. Some plastics adsorb (stick to their surface) other environmental contaminants, including pesticides and heavy metals, and may carry them into tissues where they would not otherwise go. The picture that is emerging is one of low-level, chronic exposure whose long-term health effects are genuinely not yet fully understood.

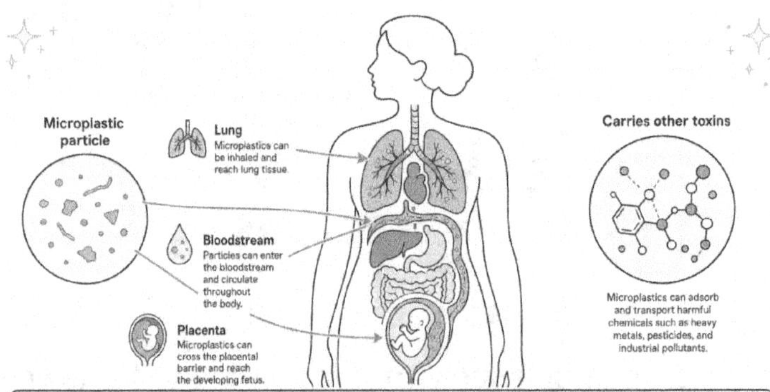

Microplastics in biological systems.

Diagram 7.10: Microplastics in biological systems

The environmental impact on marine ecosystems is better established. Fish and seabirds ingest microplastics, either directly by mistaking them for food or indirectly by eating prey that has ingested them. The particles move up the food chain. Marine invertebrates are particularly affected because they filter-feed, passing enormous volumes of water through their bodies and concentrating any particles in that water.

None of this means that every plastic in your environment is an immediate threat. The science is complex, the exposures are varied, and the dose-response relationships are not fully understood. What it means is that a material designed for its chemical inertness turns out to be present in biological systems that evolved without ever encountering it. That is a reasonable thing to pay attention to.

Bioplastics and the next generation

If the problem with conventional plastics is their chemically inert, all-carbon backbone, then the engineering solution is obvious: make a plastic whose backbone contains bonds that enzymes can attack. Build the degradation pathway into the molecule at the design stage.

This is the core idea behind bioplastics. The term is slightly ambiguous in common usage, so it is worth unpacking it. "Bioplastic" can mean two different things. It can mean a plastic made from biological feedstocks (plants rather than petroleum), a plastic designed to biodegrade, or both. These are distinct properties that do not automatically come together. You can make a biobased plastic that does not biodegrade. You can also make conventional petroleum-derived plastic that is designed to break down. The interesting category for our purposes is the biodegradable category, because it addresses the persistence problem at the molecular level.

The most commercially significant biodegradable polymer right now is PLA (polylactic acid). Its monomer is lactic acid, the same compound your muscles produce when they work without enough oxygen, the same one that makes yogurt tart. Lactic acid monomers link together through ester bonds. And ester bonds, unlike carbon-carbon bonds, are hydrolyzable. Water can break them. Enzymes can break them. Under the right conditions, PLA breaks down into lactic acid and water. Both are biologically harmless.

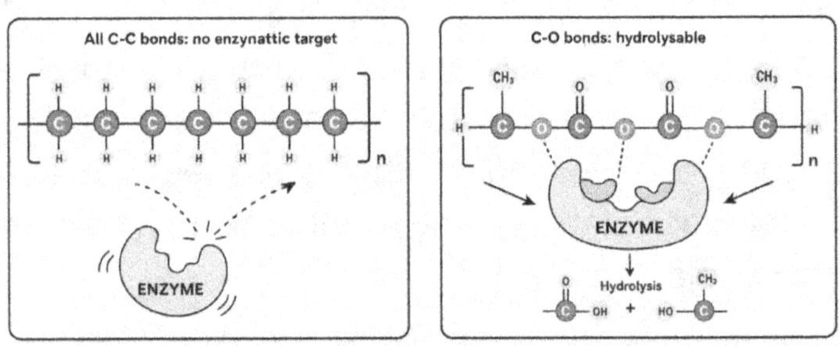

PLA vs conventional plastic backbone.

Diagram 7.11: PLA vs conventional plastic backbone

The catch, and there is always a catch, is that "biodegradable" is not the same as "compostable in your backyard." Most PLA requires industrial composting conditions to break down at a useful rate: temperatures above 55 degrees Celsius, controlled humidity, and microbial communities not present in typical landfill or ocean environments. If a PLA cup ends up in a landfill, it degrades only very slowly. If it ends up in the ocean, it behaves similarly to conventional plastic on human timescales. The promise of PLA is real, but it requires the infrastructure to match.

A second class of biodegradable polymers, the PHAs (polyhydroxyalkanoates), is more promising in this respect. PHAs are produced by bacteria as energy-storage molecules, not unlike the way your body stores energy as fat. These polymers contain ester bonds in their backbones and can be broken down by microorganisms in a wider range of environments, including marine environments. The problem is cost. Bacterial fermentation processes that produce PHA are currently much more expensive than petroleum-based plastic production. The economics are improving, but the crossover point has not yet been reached at scale.

There is also active research into enzymatically degradable synthetic polymers, building on discoveries like PETase. The idea is to engineer existing plastics to degrade faster, rather than replacing them entirely. One approach involves adding small amounts of a biodegradable co-monomer to a conventional polymer chain, creating weak points that bacteria can attack.

Another involves coating plastic products with a bacterial enzyme that activates when the plastic reaches the end of its useful life. These are early-stage ideas, but they reflect a growing recognition that the solution to the polymer problem is most likely to come from molecular engineering rather than from behavioral change alone.

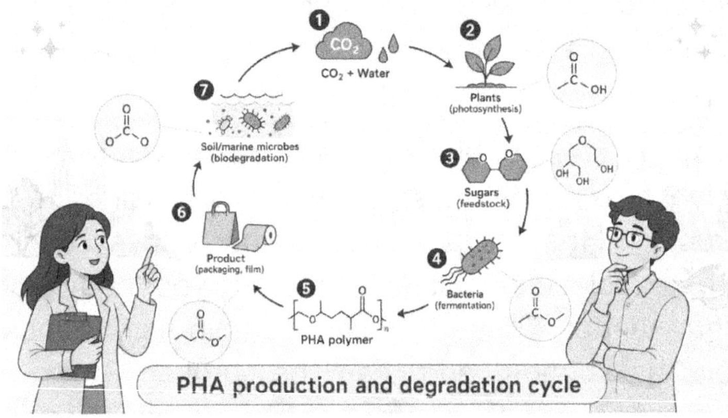

Diagram 7.12: PHA production and degradation cycle

The honest assessment is that bioplastics are not a solved problem. They are a promising research direction with real chemistry behind it, significant engineering challenges, and an infrastructure problem that no polymer chemist alone can fix. The solution to plastic pollution will require better materials, better recycling systems, better waste management, and, almost certainly, less plastic production overall. But understanding the molecular basis of the problem, understanding why a C-C backbone persists and an ester backbone does not, gives you the conceptual toolkit to evaluate claims about new materials honestly. You can now read a headline about a "fully biodegradable plastic" and ask the right question: under what conditions, and how fast?

Try This

Next time you recycle a plastic bottle, flip it over and find the resin identification code, a number inside a triangle molded into the bottom. That number maps to a specific polymer backbone. 1 is PET (polyester with ester bonds, potentially biodegradable under the right industrial conditions). 2 is HDPE (straight polyethylene chains, nearly indestructible). 5 is polypropylene. Each number is a molecular profile. Now you can read it. The number tells you what bonds hold the polymer together, which tells you what can break those bonds, which tells you whether this bottle will outlast you or not.

Takeaways

• A polymer is a long-chain molecule built by linking smaller monomers together, and the properties of the chain depend strongly on which monomer is used and how many times it repeats.

• Polyethylene, polypropylene, PVC, PET, polystyrene, and nylon each have distinct molecular personalities determined by their monomers, their chain architecture, and the types of bonds in their backbones.

• Life invented the polymer long before synthetic chemistry did: DNA is a four-monomer polymer that encodes all of heredity, proteins are twenty-monomer polymers that perform almost every chemical function in a cell, and cellulose is a glucose polymer that gives plants their structural strength.

• Conventional plastics with all-carbon backbones resist biodegradation because no enzyme has evolved to attack carbon-carbon single bonds efficiently; the same chemical stability that makes plastic useful is what makes it persist.

• Microplastics form through physical fragmentation of larger plastic items without any chemical breakdown, and they have now been detected in marine sediments, Arctic ice, and human blood and tissue.

• Biodegradable plastics like PLA and PHA solve the persistence problem by including hydrolyzable bonds, like ester bonds, that enzymes and water can attack; the current challenge is cost and the infrastructure required to ensure they actually degrade at the end of life.

• You can now evaluate claims about new plastics and biodegradable materials by asking one molecular question: what kind of bonds are in the backbone, and what can break them?

The world of plastic shows you chemistry at its most consequential: a handful of reactions, scaled up over decades, reshaping the material world and then confronting us with problems that will require the same molecular creativity to solve. The next chapter takes a gentler turn. We are going to look at the chemistry of pleasure, specifically the molecules responsible for flavor and fragrance, for the way a strawberry smells, coffee tastes, and why the right food on the right day can feel like a small act of magic. The molecules of taste and smell turn out to be some of the most elegant and personal chemistry in organic chemistry, and you are surrounded by them constantly.

8 Taste, Smell, and the Chemistry of Pleasure

You are walking down a city street, minding your own business, when it hits you. A smell. Warm yeast, caramelized crust, something sweet and brown and ancient. Before your conscious mind has a chance to weigh in, your feet have already changed direction. You are walking toward a bakery you cannot yet see, drawn by a signal older than language, more persuasive than willpower.

That signal is chemistry. Hundreds of specific molecules have just drifted off a hot loaf of bread, floated through the air, and landed on a patch of tissue the size of a postage stamp high up in your nasal passage. There, they have triggered an electrical cascade that shoots straight to the most primitive part of your brain, the part that predates rational thought, the part that says: warm food, go there, now. You did not decide to want that bread. The molecules decided for you.

Here is what you will understand by the end of this chapter: exactly which molecules were in that signal. Exactly which receptors they hit and how those receptors work. Why a baking loaf, a perfectly seared steak, and a ripe peach all produce the sensations they do. Why "natural" vanilla and "synthetic" vanilla are, at the molecular level, the same thing. And what actually happens inside a bottle of French perfume. Welcome to the most delicious chapter in organic chemistry.

How smell really works

Let's start with the most underrated sense you own. Smell is not just a pleasant add-on to your other senses. It is a direct molecular detection system, one so sensitive that your nose can detect certain compounds at concentrations of a few parts per trillion. To put that in perspective: one part per trillion is roughly one drop of water in twenty Olympic swimming pools. Your nose finds that drop.

The system works in three stages: molecule, receptor, code.

The first requirement is volatility. A molecule can only smell like anything if it can escape from its source and drift through the air to your nose. This means it has to be small (generally fewer than 300 atoms) and have enough energy to leave the liquid or solid surface it's sitting on. This is why you can smell coffee from across a room, but you cannot smell salt from across a room. Salt is an ionic crystal, effectively locked in place. Coffee contains hundreds of small, energetic, volatile organic molecules that constantly evaporate and fill the air around your cup.

Once an aroma molecule reaches your nose, it travels to the olfactory epithelium, that postage-stamp patch of tissue near the top of your nasal cavity. Embedded in this tissue are roughly 400 different types of olfactory receptor proteins. Each receptor is a protein shaped something like a twisted bundle of ribbons that threads back and forth through the membrane of a nerve cell.

The receptor has a binding pocket, a small hollow space with a specific shape and set of chemical properties. When a molecule fits into that pocket well enough, the receptor activates. The nerve cell fires. A signal goes to your brain.

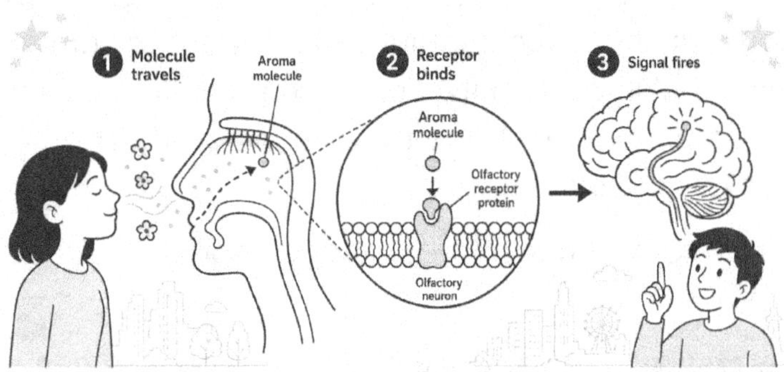

How a scent molecule triggers a nerve signal

Diagram 8.1: How a scent molecule triggers a nerve signal

Here is where it gets beautiful. The human nose does not have 400 receptors because you need 400 different receptors for 400 smells. You have 400 receptors because they work in combination, the way letters in an alphabet combine to form words.

Think about it this way. The English alphabet has 26 letters. Twenty-six symbols, but they can be arranged into hundreds of thousands of words, and those words can be arranged into every novel, poem, and news article ever written. Your olfactory system is using the same trick. A molecule of real vanilla (vanillin) might activate receptors 12, 47, 88, and 219, creating a pattern in your brain that you recognize as "vanilla." A molecule from fresh-cut grass (cis-3-hexenol) might activate receptors 3, 47, 122, and 301, a different pattern, a different word. Your brain reads the pattern, not just the individual letters. Four hundred receptor types, combined in various ways, can theoretically distinguish trillions of odors. The human nose is, by any reasonable standard, a combinatorial supercomputer.

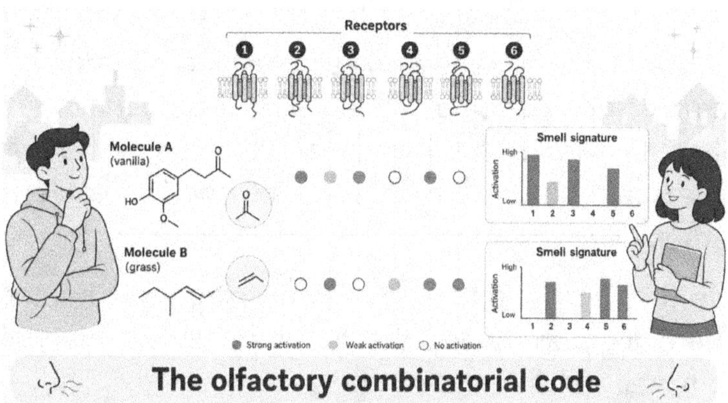

Diagram 8.2: The olfactory combinatorial code

Now, smell and taste are not the same sense, but they conspire. Taste receptors on your tongue detect only five basic categories: sweet, salty, sour, bitter, and umami (savory). That's it. Five channels of information. So how does food seem to have infinite variety? Because what you experience as "flavor" is almost entirely the result of aroma molecules traveling from the back of your mouth up behind your palate to your olfactory epithelium. This is called retronasal smell. When you have a bad cold, and your nose is blocked, food tastes like nothing, not because your tongue stops working, but because retronasal smell is cut off. The five basic tastes remain. Everything else goes silent.

This means that most of what you think of as taste is actually smell, processed from behind. The molecular richness you experience in a cup of coffee (over 800 distinct volatile compounds have been identified in roasted coffee) is reaching your brain almost entirely through your nose, even as you drink.

The major aroma families

Nature did not randomly scatter aroma molecules across the periodic table. The vast majority of interesting food and fragrance smells come from four molecular families, each with its own characteristic personality.

The first family: esters. You met esters in Chapter 3, where they explained the fruity smells of everything from nail polish remover to banana candy. They are formed when an alcohol and a carboxylic acid link up and lose a water molecule. What you may not remember is that esters are almost universally fruity. Ethyl acetate smells like nail polish remover (and also contributes to the aroma of wine). Isoamyl acetate is the dominant molecule in banana candy. Ethyl butyrate is the chief character in pineapple. If you ever wondered why cheap candy smells so intensely fruity, it is because a single well-chosen ester can telegraph "tropical fruit" with startling efficiency. Real fruit is more complex, but the ester sets the tone.

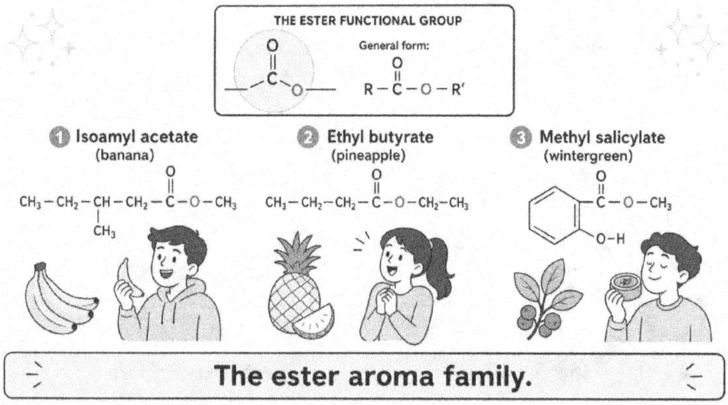

Diagram 8.3: The ester aroma family

The second family: terpenes. These molecules are built from a repeating five-carbon unit called isoprene and are found throughout the plant world. Limonene smells like orange peel; it is the molecule you release when you zest a lemon, the one that mists into the air when you peel a clementine. Pinene is the smell of pine forests. Linalool is the lavender molecule. Menthol is the cooling sensation of peppermint, which works not only by smell but by physically activating cold-sensing receptors on your skin, tricking your nervous system into perceiving coolness that isn't there.

Terpenes are also the molecules responsible for the smell of cannabis (a family of terpenes called myrcene, limonene, and caryophyllene), the resinous quality of gin (juniper terpenes), and the sharp brightness of fresh herbs. When you smell a rosemary sprig, you are smelling terpenes that the plant synthesizes as insect repellents. The forest does not smell like the forest for your benefit. It smells that way because it is trying to deter predators.

The terpene family: nature's volatile defenders.

Diagram 8.4: The terpene family: nature's volatile defenders

The third family: aldehydes. Aldehydes are organic compounds with a very specific functional group at the end of their carbon chain: a carbon double-bonded to oxygen and single-bonded to a hydrogen. Small aldehydes (like acetaldehyde, two carbons) are sharp, almost harsh, and contribute to the sting of vinegar and the edge of white wine. But as aldehydes get larger, something interesting happens. Medium-chain aldehydes (around eight to twelve carbons) become distinctly floral, which is why aldehydes are a signature ingredient in classic perfumes. Very short-chain aldehydes are pungent; very long-chain aldehydes are almost odorless. The chain length is everything.

Benzaldehyde is the almond smell, the smell of marzipan, and cherry pits (the pits contain small amounts of benzaldehyde, which is why they smell vaguely like artificial almond extract). Cinnamaldehyde is, as the name suggests, cinnamon. And vanillin, the most important aldehyde in the food world, we will address in detail very shortly.

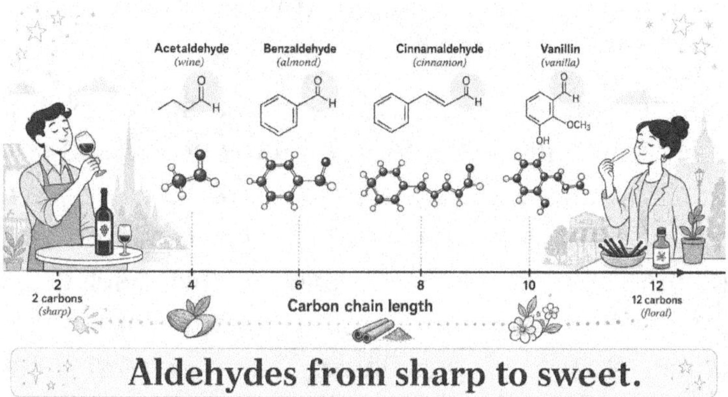

Diagram 8.5: Aldehydes from sharp to sweet

The fourth family: sulfur compounds. Here we get into the most divisive corner of the aroma world. Sulfur is an element that, in small organic molecules, produces smells of stunning intensity and complexity. This is partly chemistry and partly evolution: you are wired to pay attention to sulfur smells because many toxic or spoiled things smell sulfurous, but also because many deeply nutritious things do.

The sulfur compounds in garlic (allicin and its breakdown products) are activated when you crush a clove. The garlic cells rupture, enzymes mix with previously separated compounds, and within seconds, a burst of volatile sulfur molecules fills the air. Onion does the same thing, with a slightly different set of sulfur compounds, including propanethial S-oxide, which is also responsible for making you cry (it reacts with moisture in your eyes to form a mild sulfuric acid).

The molecules that give cooked meat its savory smell are partly sulfur-based. The molecule that gives a hard-boiled egg its distinctive smell (hydrogen sulfide) is just two atoms: sulfur and hydrogen.

And then there is durian. Durian, the spiky Southeast Asian fruit that is banned from many hotels and public transit systems, is a masterclass in sulfur complexity. Durian contains a cocktail of sulfur compounds unlike anything else in the fruit world: diethyl disulfide (the smell often compared to rotten onions), propanethiol (which smells of gas leaks), and several others. Layered over this is a sweet, creamy, floral base of esters and other fruity molecules. The result is a fruit that smells simultaneously like a blocked drain and tropical custard. People either find it transcendent or unlivable, often depending on whether they encountered it early in life. The durian teaches a deep lesson: smell is not inherently pleasant or unpleasant. It is contextual, combinatorial, and shaped by experience.

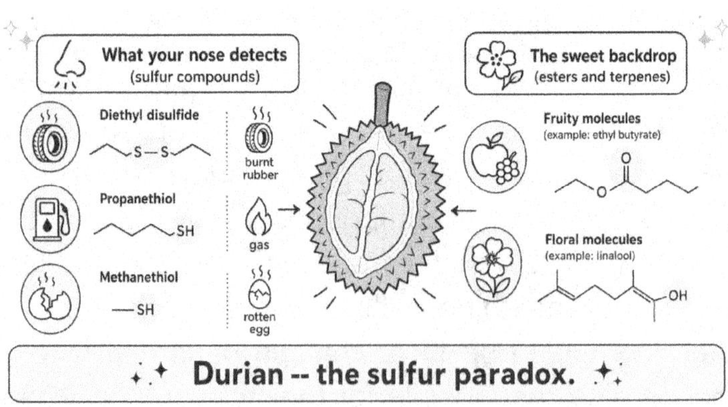

Diagram 8.6: Durian, the sulfur paradox

The "natural" vs "synthetic" myth

Let's talk about vanilla, because vanilla is the most instructive argument in the entire flavor industry.

The vanilla plant (Vanilla planifolia) is a flowering orchid native to Mexico. It produces seed pods, vanilla beans, that contain hundreds of flavor compounds. The most important of these, the molecule that your brain has learned to recognize as "vanilla," is vanillin: a simple aromatic aldehyde with two six-carbon ring features and a handful of attached groups.

Vanilla beans are expensive. Growing them requires hand pollination (the natural pollinators are specific bees found only in the plant's native range), an 18-month maturation period, careful curing, and significant labor. Real vanilla extract, therefore, costs real money.

Vanillin, the molecule, can also be synthesized from other sources. The most common industrial route uses lignin, the structural polymer from wood pulp (a byproduct of the paper industry). Lignin is chemically transformed into vanillin through a series of steps. The resulting molecule is identical, atom for atom and bond for bond, to the vanillin extracted from a vanilla bean.

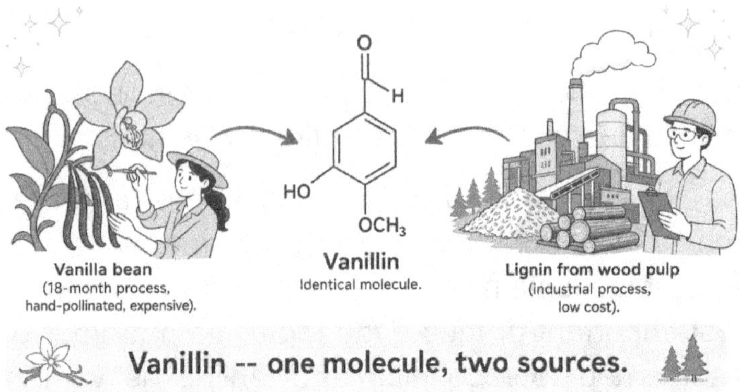

Diagram 8.7: Vanillin, one molecule, two sources

This is not a controversy in chemistry. The molecules are the same. A spectroscopist with the most sensitive instruments in the world cannot tell you, from the vanillin molecule alone, whether it came from a bean or a factory. They would need to look at isotope ratios (the proportions of carbon-13 vs carbon-12 and other trace signatures from the growing environment), and even then, it is only statistical inference, not a categorical difference.

So why does the label say "natural vanilla flavor" on one product and "artificial vanilla flavor" on another? Regulatory definitions, not molecular reality. "Natural" means the flavor compound was derived from a natural source (plant, animal, fermentation). "Artificial" means it was produced through chemical synthesis. Same molecule. Different origin story. Different label. Different price.

Real vanilla extract does contain additional compounds beyond vanillin, p-hydroxybenzaldehyde, 4-

methylguaiacol, and dozens of others that contribute to its complexity. So pure vanillin, natural or synthetic, does not taste exactly like real vanilla extract. It tastes like the main character, without the supporting cast. But the "natural" label does not guarantee this complexity; natural vanillin derived from lignin is still just vanillin, just like the synthetic version. The distinction the label is drawing is not "complex vs simple" or "better vs worse." It is purely about where the molecule was made.

This matters beyond vanilla. The entire $20 billion flavor industry runs on this logic. Isoamyl acetate (banana flavor) can be produced by fermenting certain yeasts. Ethyl butyrate (pineapple) can be made the same way. These fermentation-derived molecules carry the "natural" label. Their synthetically identical cousins, made in a lab, carry the "artificial" label. The molecules do not know the difference. Your tongue does not know the difference. Only the label knows the difference.

The Maillard reaction: why browning tastes good

At this point, you know about volatile aroma molecules and receptor codes. Now we need to ask a more specific question: where do all these molecules come from during cooking? The answer is not one reaction. It is hundreds of reactions happening simultaneously in a controlled molecular chaos. But the most important one has a name, and you should know it.

In 1912, a French physician named Louis-Camille Maillard mixed amino acids and sugars and heated them. What he got was a brown, complex, extraordinarily fragrant mixture. He published his findings in a chemistry journal, and for decades, food scientists and chefs had no idea. The "Maillard reaction", technically, a family of related reactions, is now understood to be responsible for the flavors of browned meat, toast, roasted coffee, beer malt, baked bread crust, seared vegetables, chocolate, and caramelized onions.

Here is what happens. When you heat a food containing both amino acids (from proteins) and reducing sugars (from carbohydrates) to above roughly 140 degrees Celsius, the amino group of the amino acid attacks the carbonyl group of the sugar. This initial pairing begins a cascade of rearrangements, fragmentations, and recombinations. It is not a single reaction; it is a chain reaction with hundreds of forks in the road, and each fork produces different flavor molecules.

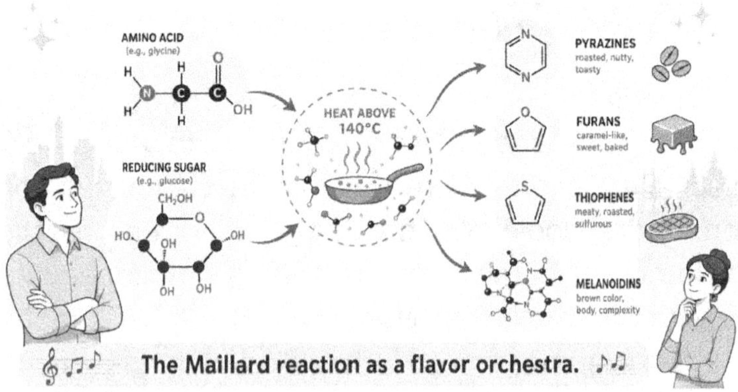

The Maillard reaction as a flavor orchestra.

Diagram 8.8: The Maillard reaction as a flavor orchestra

The result of all those forks is a collection of molecules called pyrazines (roasted, nutty, the dominant class in coffee and chocolate), furans (sweet, caramel-like), and melanoidins (the large brown polymers that give browned food its color). A perfectly seared steak has over 600 distinct volatile compounds in its crust, almost all of them Maillard products, and almost none of them are present in raw beef.

This is the molecular orchestra analogy. Raw ingredients are the individual instruments sitting silent in a pit. Heat is the conductor raising the baton. And the Maillard reaction is not one note; it is a full symphony, each section playing simultaneously, the harmonics of one instrument shaping the timbre of the next. The difference between a pale, steamed piece of chicken and a golden-crusted, beautifully seared one is not about crispness or appearance. It is about the presence or absence of six hundred flavor molecules that do not exist until you apply enough heat to launch the Maillard cascade.

This is also why chefs insist on high heat for searing. A pan that is not hot enough will not reach the Maillard threshold. The food will steam in its own moisture, and the proteins and sugars will sit there, unreacted. You will get tender food that tastes like nothing much. This is not a matter of preference. It is chemistry.

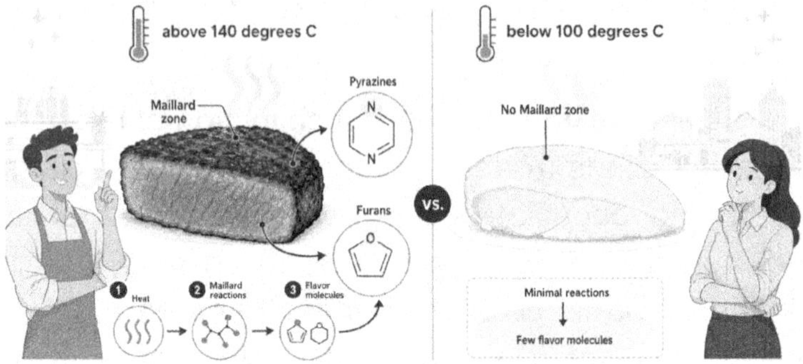

Seared vs. steamed -- a Maillard comparison.

Diagram 8.9: Seared vs. steamed, a Maillard comparison

Caramelization is related but distinct. When you heat pure sugar to high temperatures (around 160-180 degrees Celsius) without any amino acids present, the sugar molecules break apart and recombine in a different set of reactions, producing a different family of flavor compounds, including diacetyl (buttery), furans, and lactones. Caramelization is happening in your caramel sauce. The Maillard reaction is happening on your steak. Both are happening simultaneously in roasted coffee and baked bread, which is part of why those foods have such layered complexity. They are running two flavor-generating programs simultaneously.

Fermentation: yeast turns sugar into flavor

Beer, wine, bread, cheese, sourdough, vinegar, miso, kimchi, the fermented foods are arguably humanity's greatest accidental chemistry experiment. And they are all, at their core, the work of microorganisms doing what microorganisms have always done: eating sugar and excreting products.

The core reaction of fermentation is well known: glucose (a six-carbon sugar) is consumed by yeast, which produces ethanol (a two-carbon alcohol) and carbon dioxide. That's the headline. Bread rises because of the carbon dioxide bubbles. Beer and wine contain ethanol. But the headline misses the story.

Fermentation is not a single clean reaction. Yeast metabolism is a complex network of enzymatic reactions, and dozens of these reactions produce flavor compounds as byproducts. Ethyl acetate (the ester responsible for a fruity, nail-polish edge in some wines and ales) is a fermentation byproduct. Diacetyl (the molecule that tastes buttery, the one that brewers try hard to eliminate from beer by giving yeast time to reabsorb it) is produced by yeast during fermentation. Isoamyl acetate, the banana ester, is present in higher concentrations in certain ale yeasts and is a deliberate feature of hefeweizen (wheat beer).

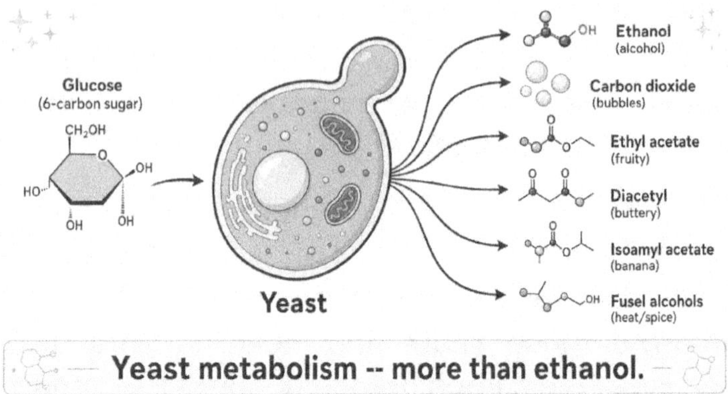

Diagram 8.10: Yeast metabolism, more than ethanol

Different yeast strains produce different ratios of these flavor molecules, which is why Belgian ale tastes completely different from English bitter, even if both are made from similar base grains. The yeast is the chef. The grain is the ingredient. The fermentation conditions, temperature, time, and oxygen level are the recipe.

Sourdough bread adds a further layer: bacterial fermentation. Wild sourdough starters contain not just yeast but lactic acid bacteria (Lactobacillus species), which produce lactic acid and acetic acid (vinegar) as fermentation products. These acids give sourdough its tang, lower the dough's pH, and modify gluten structure in ways that change the texture. They also generate their own suite of flavor esters. A long, slow, cold fermentation allows more time for these bacterial flavors to develop, which is why a sourdough loaf left to proof overnight in the refrigerator tastes more complex than one rushed through a warm-room rise.

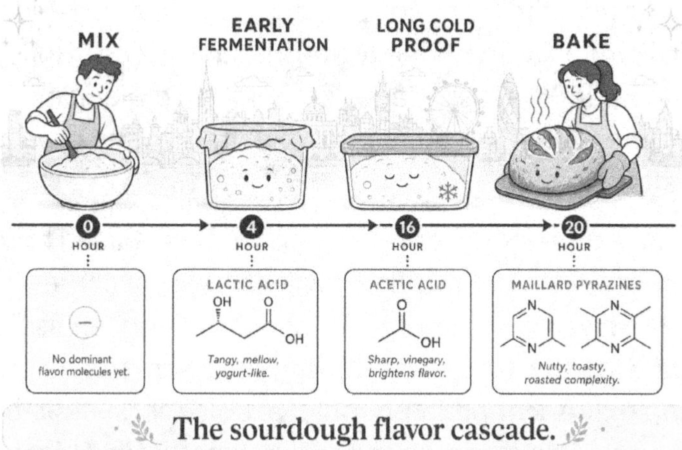

Diagram 8.11: The sourdough flavor cascade

Fermentation also explains the flavors of vinegar, miso, soy sauce, fish sauce, and aged cheese, all of which are microbially driven flavor transformations. Fish sauce, made from fermented fish and salt, contains hundreds of flavor compounds, including glutamates (umami intensity), a complex blend of fermentation esters, and controlled amounts of volatile sulfur compounds that, at the right concentration, read as "savory depth" rather than "something has gone wrong." The difference between fish sauce tasting wonderful and fish sauce tasting terrible is, again, about concentration and context. Molecules do not have inherent preferences. Your brain has learned associations.

Inside a perfume bottle

The flavor world and the fragrance world share most of their molecules, but they pursue them differently. A food chemist is trying to recreate an edible experience. A perfumer is composing an emotional one. Both are working in the language of volatile molecules. But perfumers have an additional constraint: their creations must evolve.

When you apply a perfume, you notice this. There is what you smell in the first few minutes, which may be bright, citrusy, or sharp. Then something shifts. The middle of the fragrance emerges, warmer and more complex. An hour later, only a soft, persistent base remains. This is not your nose getting tired. This is the perfume pyramid.

The pyramid works because different molecules have different vapor pressures, different eagerness to evaporate. Top notes are the most volatile molecules: small, light, quick to evaporate. Think citrus terpenes (limonene, bergaptene), certain fresh aldehydes, and simple herbal molecules. They hit your nose immediately and are mostly gone within 20-30 minutes. Heart notes (also called middle notes) are moderately volatile: floral molecules like linalool (lavender), geraniol (rose), and jasmine-derived compounds. They emerge as the top notes fade and persist for an hour or two. Base notes are the least volatile molecules: large, heavy, persistent. Musks (originally derived from animal secretions, now almost always synthetic), sandalwood-derived sesquiterpenes, vanilla-family molecules, resins, and ambers. These anchor the fragrance, fixing the lighter molecules against your skin and slowing their evaporation. They may persist for hours or even days.

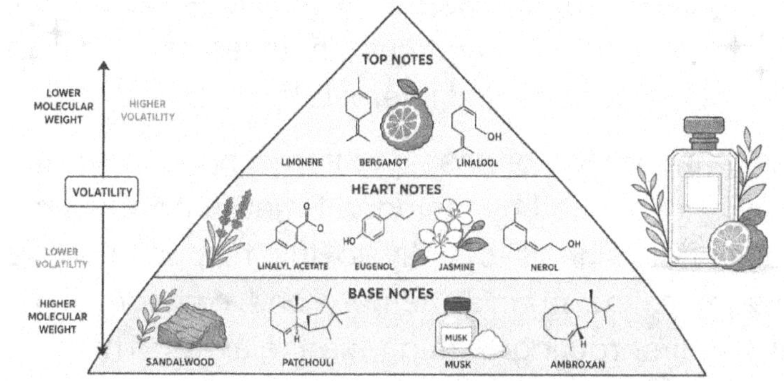

The perfume pyramid.

Diagram 8.12: The perfume pyramid

A master perfumer, called a "nose" in the industry, can identify several thousand distinct aroma molecules by smell alone. Professional training takes years, and even then, the skill is partly innate. But the science underneath it is the same science that explains why bread smells good: volatile molecules, receptor activation, combinatorial code, and a brain that has spent a lifetime building associations.

The fragrance industry synthesizes the vast majority of its aroma molecules. Only a small fraction of fine fragrances use genuinely natural extracts, such as rose absolute, sandalwood oil, and ambergris (now nearly always replaced with synthetic musks for conservation reasons). And this is not a compromise. Many of the most celebrated perfumes in history were built around synthetic molecules unavailable in nature. Calone smells of ozone and ocean spray. Iso E Super is a molecule with no natural equivalent and is found in hundreds of iconic woody fragrances. The perfumer's palette is not constrained by what plants produce. It is constrained only by what human noses can detect and what human brains find interesting.

And what human brains find interesting is, it turns out, heavily influenced by memory. Smell is the only sense that sends signals directly to the hippocampus and amygdala, the brain's memory and emotion centers, without first routing through the thalamus. This is why smells trigger involuntary, emotionally loaded memories in a way that sight and sound rarely do. It is why the smell of sunscreen transports you to a specific beach at a specific age. The molecules are not just delivering information. They are unlocking files.

Try this

Next time you cook something with a crust, a steak, a piece of toast, or a baked potato, smell the raw ingredient before it goes into the heat, and then smell it again the moment it starts to brown. You are tracking the Maillard reaction in real time. The first smell is simple, almost plain. The second smell is dozens of newly-formed pyrazines and furans arriving in your nose simultaneously, molecules that did not exist five minutes ago. You are smelling chemistry happening.

Takeaways

• You can explain why smell works as a combinatorial code: roughly 400 receptor types, each activated by different molecules, produce patterns that your brain reads as distinct smells, the same way 26 letters produce thousands of words.

• You can identify the four major aroma families: esters (fruity), terpenes (citrus, pine, herbal), aldehydes (almond, vanilla, floral), and sulfur compounds (onion, garlic, savory depth), and describe the molecular feature that unifies each family.

• You can explain, precisely, why "natural" and "synthetic" vanillin are chemically identical, and what the "natural" label on a food product actually means (origin of synthesis, not molecular difference).

• You can describe the Maillard reaction: the combination of amino acids and reducing sugars under heat that produces hundreds of flavor molecules and is responsible for the flavor of seared meat, toast, roasted coffee, baked bread crust, and chocolate.

• You can explain fermentation as a flavor-generating process, not just an alcohol-producing one. Different yeast strains and bacterial partners produce different flavor esters, acids, and other volatiles that define the character of beer, wine, bread, and fermented foods.

• You can describe the perfume pyramid (top, heart, and base notes) in terms of molecular weight and vapor pressure, and explain why a fragrance evolves as lighter molecules evaporate before heavier ones.

• You can explain why smell triggers emotional memories more powerfully than any other sense, because the olfactory signal bypasses the thalamus and goes directly to the brain's memory and emotion centers.

The chemistry of pleasure is not soft science. It is specific molecules, specific receptors, specific cascades of reactions controlled by temperature and time. Every smell you have ever loved was a molecular event. Every flavor that has ever surprised you was a reaction you did not know was running.

Now that you can read the molecular language of taste and smell, there is one more thing molecules do that shapes the entire modern world. They burn. They release energy. They move cars, heat homes, and power the civilization you live in. In Chapter 9, we are going to take the carbon chemistry you know and point it directly at the question of energy: where it comes from, why fossil fuels work the way they do, and what it will take to replace them.

9 Burning and Beyond: Fuels and Energy

You pull into the gas station. The nozzle clicks into the tank. The numbers spin on the pump. Ten gallons. Forty dollars. A mundane Tuesday errand.

But stop for a second and think about what you just bought.

Three hundred million years ago, dense mats of marine algae and microscopic organisms lived in shallow tropical seas. They pulled carbon dioxide out of the ancient air. They stitched those carbon atoms into long, energy-rich chains using nothing but sunlight and seawater. Then they died. Over millions of years, buried under layer after layer of sediment, subjected to heat and pressure that would crush a submarine, those molecules were transformed. Carbon chains were rearranged, cracked, and concentrated into the substance we now call crude oil.

You just paid forty dollars for ten gallons of ancient sunshine.

That sunshine traveled 93 million miles to reach those algae. It was captured in a chemical bond, a carbon-hydrogen bond, and it has been sitting in the dark underground, perfectly preserved, ever since. Today, it gets pumped into your tank, and within the hour, it will be driving a series of controlled explosions in your engine, releasing that same solar energy as motion and heat.

This chapter is about that journey. Energy in, energy out. Combustion and its gentler cousin, respiration. The plants that capture light and the animals that spend it. And the enormous challenge of finding something, anything, that can carry as much energy as a gallon of gasoline without costing us the climate in the process.

By the time you finish this chapter, you will be able to explain what is happening, at the molecular level, every time someone fills a tank, eats a meal, or charges an electric car. You will understand why replacing gasoline is genuinely hard, not a failure of political will, but a problem written in chemistry. And you will understand why the answer, when we find it, will also be written in chemistry.

What combustion really is

Here is what most people think combustion is: fire.

Fire is the visible part: the light and heat. But combustion is happening before you can see any flame, and it is happening at a molecular level in a way that is far more interesting than simple burning.

Let's start with what you are actually burning. Gasoline is not one thing. It is a mixture of hydrocarbon molecules, mostly chains of six to twelve carbon atoms with hydrogen atoms attached along every available position. Think of them as short, fat molecular centipedes. Carbon spine, hydrogen legs. These are the same kinds of molecules you met earlier in this book, the alkanes, the simplest family of organic molecules.

Now, carbon-hydrogen bonds store energy. Not because there is something magical going on, but because the carbon and hydrogen atoms are held apart from each other at a very specific distance by a very specific amount of force. That force has a direction and a magnitude, and that magnitude represents stored potential energy, like a compressed spring.

The same is true of carbon-carbon bonds. Every single C-H and C-C bond in a hydrocarbon molecule is a tiny energy storage unit.

When you combust a hydrocarbon, here is what happens at the molecular level. Oxygen molecules collide with the hydrocarbon molecules. With enough energy to get started, more on that in a moment, bonds begin to break. The carbon-hydrogen bonds snap. The carbon-carbon bonds snap. The atoms are freed. Then those atoms form new bonds with oxygen. Carbon binds to two oxygen atoms to form carbon dioxide. Hydrogen binds to one oxygen atom to form water.

Both of those new bonds, C to O in carbon dioxide, H to O in water, are stronger and lower-energy than the original C-H and C-C bonds. Lower energy means more stability. The difference between the energy stored in the original bonds and the energy stored in the new bonds has to go somewhere.

It goes into your engine. Into the air. Into your hand when you hold it near a candle.

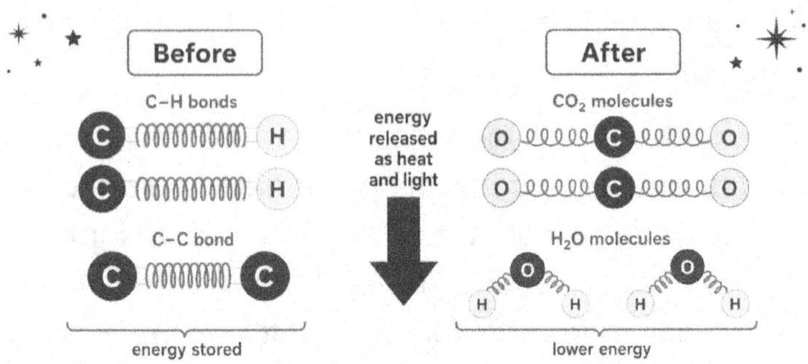

Bond energy bookkeeping in combustion.

Diagram 9.1: Bond energy bookkeeping in combustion

The energy released is called the enthalpy of combustion. This is a thermodynamics term, and if you want to go deep on enthalpy, our sibling volume in this series on thermodynamics is exactly the place to do it. For now, think of it simply as the total energy released when you burn a given amount of fuel. It is a property of the molecule. It is fixed. And it is why octane (a specific eight-carbon chain in gasoline) releases a very specific, predictable amount of energy every single time it combusts.

Here is the part that surprises most people. Combustion does not just happen spontaneously at room temperature, even though it would release energy. You need something to get it started. You need a spark, or a flame, or enough heat. You already know this concept from Chapter 5: the activation energy. It is the push that sets the molecular dominoes in motion. Once combustion is underway, it releases enough heat to ignite neighboring molecules, sustaining the reaction. But without that initial push, gasoline sits in your tank, harmlessly, all winter.

Activation energy is your engine's spark plug's entire reason for existing. One tiny electrical spark provides just enough energy to break the first few bonds. The heat from those bonds ignites more molecules. Within milliseconds, you have a controlled explosion powerful enough to push a piston down a cylinder.

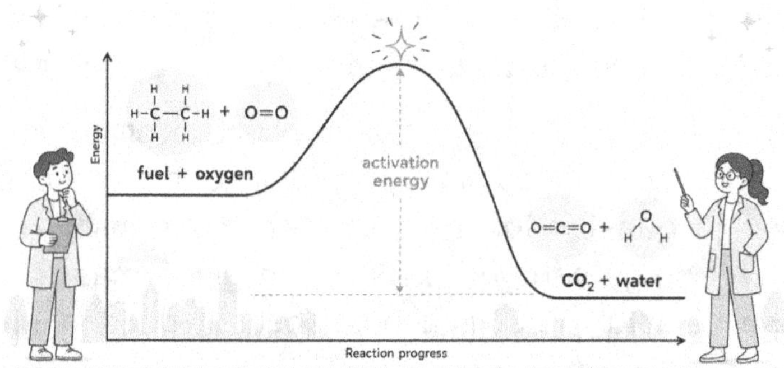

Activation energy and combustion

Diagram 9.2: Activation energy and combustion

There is one more thing worth saying about combustion before we move on. Complete combustion, where every carbon atom ends up in carbon dioxide and every hydrogen ends up in water, is actually the ideal case. In a real engine, or a real fireplace, or a real candle, combustion is never perfectly complete. Some carbon atoms form carbon monoxide instead of carbon dioxide. Some form tiny soot particles, small enough to be invisible individually but collectively responsible for the black residue on a chimney or the haze over a city. Your catalytic converter exists specifically to mop up these incomplete combustion products and push them toward the complete reaction.

The chemistry of combustion is elegant. The engineering of doing it cleanly is where the hard work is.

Energy density: why gasoline is hard to replace

Now, let's talk about why replacing gasoline is one of the hardest chemistry problems of the 21st century.

It is not a question of political courage or corporate greed (though those play roles). It is, at a fundamental level, a question of molecular energy density. How much energy can you pack into a given mass? And gasoline, it turns out, is almost absurdly good at this.

A kilogram of gasoline stores about 44 megajoules of energy. That is not a number you need to memorize; get a feel for it. A megajoule is roughly the energy in a quarter-pound hamburger. Forty-four of those are packed into one kilogram of liquid that you can pump in 90 seconds and carry in a sealed tank without any special equipment.

Ethanol, which is the primary biofuel additive in gasoline sold in the United States, stores about 26 megajoules per kilogram. That is roughly 60 percent of what gasoline stores. You need 40 percent more ethanol to go the same distance. That is why cars running on high-ethanol blends get slightly worse gas mileage.

Why the difference? Chemistry again. Ethanol has the formula of a two-carbon chain with an attached oxygen atom. That oxygen atom is already somewhat oxidized; it has already given up some of its energy potential by bonding to oxygen. The C-H and C-C bonds in pure hydrocarbons have never done that. They are fully reduced, maximally energy-rich.

Now here is where it gets really interesting. Hydrogen gas, pure H2, no carbon at all, stores about 120 megajoules per kilogram. That is roughly three times the energy of gasoline, per kilogram. If energy density were the only thing that mattered, we would all be driving hydrogen cars right now.

The reason we are not comes down to a single, brutal physical reality. Hydrogen molecules are tiny. They are so small that they slip between the atoms of metal containers. Storing hydrogen as a gas requires enormous pressurized tanks. Storing it as a liquid requires cooling it to minus 253 degrees Celsius, which is just 20 degrees above absolute zero. These are engineering problems on a scale that makes ordinary fuel logistics look simple.

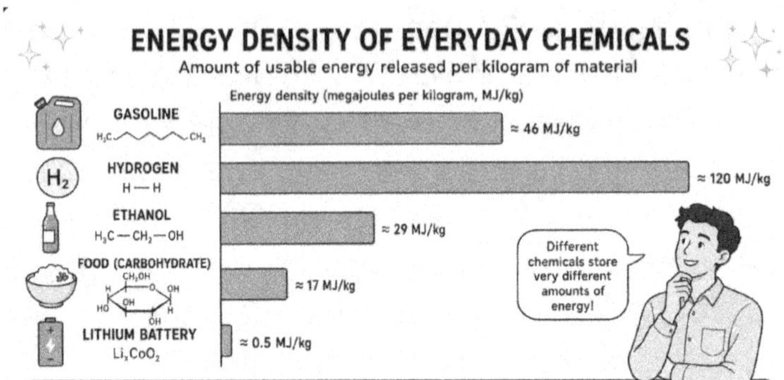

Energy density comparison.

Diagram 9.3: Energy density comparison

Lithium-ion batteries, which power most electric cars, store roughly 0.7 to 0.9 megajoules per kilogram. That is about 2 percent of gasoline's energy density by mass. But wait, electric motors are also about three to four times more efficient than combustion engines. So the real comparison is closer to 6 or 8 percent. That is still a massive difference, which is why the battery pack in an electric car weighs 400 to 600 kilograms, while a full tank of gasoline weighs about 25 to 30 kilograms.

For comparison, a kilogram of carbohydrates, the kind of fuel your body runs on, stores about 17 megajoules. Fats store about 37 megajoules per kilogram, which is why your body prefers to store long-term energy as fat rather than as starch. Your body is doing its own version of fuel optimization.

The point of all these numbers is not to make you despair about electric vehicles or hydrogen. It is to give you a genuine appreciation of why the energy transition is a hard problem. We are not replacing a mediocre technology. We are trying to replace one of the most energy-dense, conveniently stored, easily transported substances chemistry has ever produced. The answer will require not just engineering, but new chemistry, new molecules, and new ways of storing energy that we may not have fully imagined yet.

Photosynthesis: the original solar cell

Before we talk about where fossil fuels come from, let's talk about where all biological energy comes from in the first place.

Plants figured out how to run on sunlight around 3.5 billion years ago. The mechanism they use, photosynthesis, is so elegant that engineers are still trying to copy it. And the product they make, glucose, a simple six-carbon sugar, is the starting point for almost every energy-containing organic molecule on Earth.

Here is the basic idea. A leaf is a solar panel. The green color you see is from a pigment molecule called chlorophyll, and chlorophyll's job is to absorb photons, specifically the red and blue wavelengths of light, which is why leaves reflect the green. When a chlorophyll molecule absorbs a photon, it gets energized, like a compressed spring. That energy is then used to drive chemical reactions that would otherwise not occur spontaneously.

That chemistry involves carbon dioxide and water. The plant pulls carbon dioxide out of the air through tiny pores in its leaves. It pulls water up from the soil through its roots. And using the energy from those captured photons, it breaks apart those molecules and reassembles the carbon, hydrogen, and oxygen atoms into glucose.

Glucose is a six-carbon molecule with a carefully arranged pattern of carbon-hydrogen and carbon-oxygen bonds. Some of those bonds store significant energy. The plant has essentially captured energy from photons and locked it in chemical bonds, converting one form of energy into another. This is the original solar cell, not silicon and semiconductors, but chlorophyll and enzymatic chemistry.

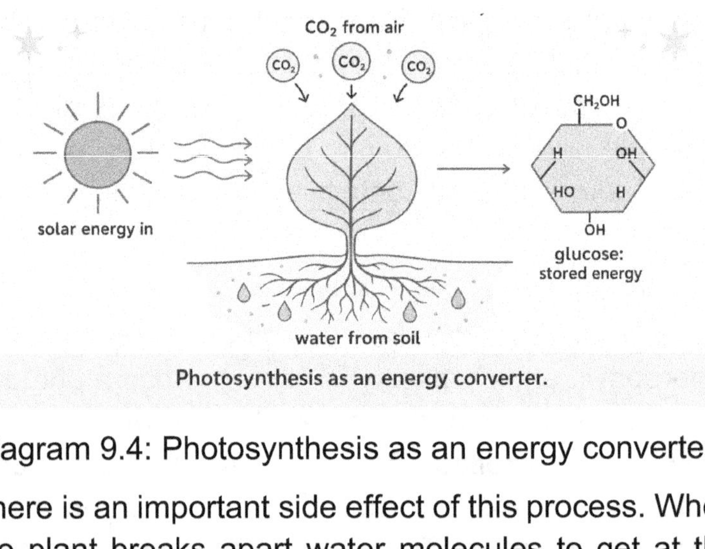

Photosynthesis as an energy converter.

Diagram 9.4: Photosynthesis as an energy converter

There is an important side effect of this process. When the plant breaks apart water molecules to get at the hydrogen atoms it needs, it releases oxygen as a byproduct. Every molecule of oxygen in the air you are breathing right now was produced by photosynthesis. Plants did not set out to create a breathable atmosphere; they were building glucose and discarding the oxygen they did not need. The fact that we evolved to run on that oxygen is one of the more remarkable accidents in the history of life on Earth.

Now, glucose does not just stay in the plant. Plants use it for energy when they need to grow or reproduce. They convert it into cellulose to build their cell walls. They convert it into starch to store energy for later. Animals eat plants and extract the glucose energy for themselves. And over geological timescales, when organisms die and are buried without decomposing, that organic matter, with all its stored glucose-derived energy, becomes the raw material for fossil fuels.

This is why we say gasoline is ancient sunshine. It is not a metaphor. It is a literal description of the energy pathway. Photons hit ancient plants. Those plants built glucose. The glucose was transformed, over hundreds of millions of years, into the hydrocarbons in crude oil. You then combust those hydrocarbons and release the same photons' worth of energy as motion in your car.

The bad news is that this process took 300 million years, and we are reversing it in about 300 years.

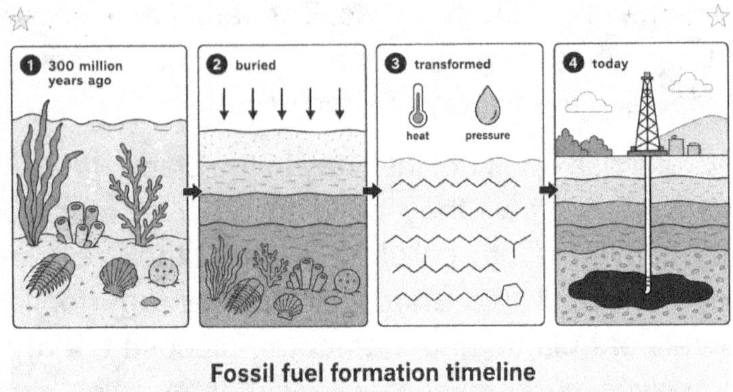

Diagram 9.5: Fossil fuel formation timeline

Cellular respiration: combustion done gently

Here is a question that should bother you: if combustion releases energy, why don't you burst into flames when you eat a sandwich?

The answer is that your cells are doing something remarkable. They are running what is essentially the same chemical process as combustion, breaking C-H bonds and ultimately producing carbon dioxide and water, but doing it in hundreds of tiny controlled steps, each carefully managed by enzymes, releasing the energy in small, useful parcels rather than all at once.

This process is called cellular respiration, and it is one of the most impressive pieces of biochemical engineering in existence.

Let's start with the fuel. You eat glucose in your sandwich bread, in the sugars in your food, in almost everything carbohydrate-based. Your digestive system breaks down complex sugars and starches into individual glucose molecules, which are then absorbed into your bloodstream. When those glucose molecules reach your cells, the real work begins.

The first stage of cellular respiration happens in the fluid inside the cell. Glucose is broken apart, step by step, into smaller molecules. This releases a little energy. But the bigger energy payoff comes in the second and third stages, which occur inside specialized cellular compartments called mitochondria.

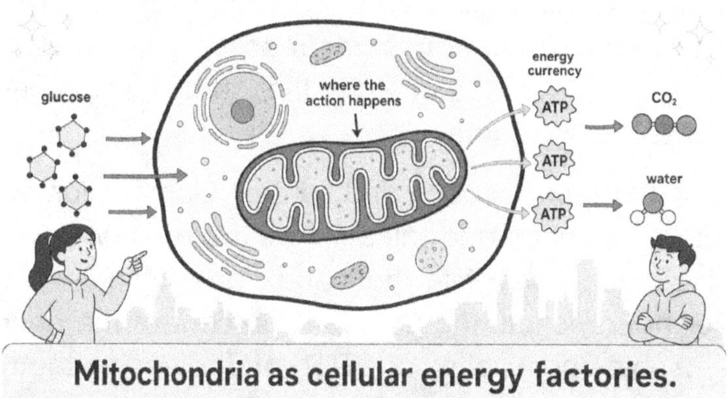

Mitochondria as cellular energy factories.

Diagram 9.6: Mitochondria as cellular energy factories

Inside the mitochondria, glucose fragments are further broken down, and the energy is captured in a specific form. Think of it as chemical currency. The energy released from breaking C-H and C-C bonds is used to attach a phosphate group onto a molecule called adenosine diphosphate, creating adenosine triphosphate, which almost everyone knows by the initials ATP.

ATP is the energy currency of your body. Every time a muscle fiber contracts, it spends ATP. Every time a neuron fires a signal, it spends ATP. Every time a cell builds a protein, it spends ATP. Your body produces and consumes roughly your body weight in ATP every single day, a cycle of constant molecular energy exchange.

The final products of combustion are the same: carbon dioxide and water. You exhale the carbon dioxide. The water is used in various ways or excreted. From a purely molecular bookkeeping standpoint, respiration and combustion are the same reaction. The inputs are the same. The outputs are the same. The energy released is the same.

The difference is speed and control. Combustion does it in milliseconds and releases all the energy as heat. Respiration does it in thousands of enzymatic steps over hours, capturing about 40 percent of the energy as usable chemical work (ATP) and releasing the rest as body heat. That 37 degrees of warmth you carry around with you? That is your respiration inefficiency, and it is what keeps you alive in winter.

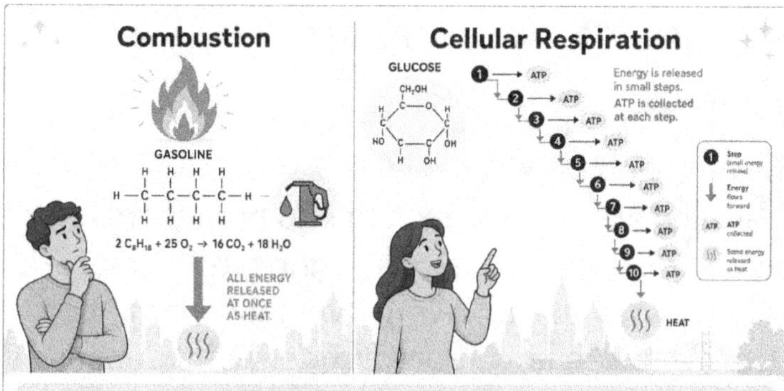

Combustion vs. cellular respiration.

Diagram 9.7: Combustion vs. cellular respiration

There is a beautiful symmetry here that is worth sitting with. Photosynthesis uses solar energy to convert carbon dioxide and water into glucose. Cellular respiration takes glucose and, using oxygen, breaks it back into carbon dioxide and water while releasing the same energy. These are the reverse of each other, almost perfectly. One process stores energy in bonds. The other releases it.

If you drew the carbon cycle, the movement of carbon atoms through air, plants, animals, and back to air, you would see that photosynthesis and respiration together are its engine. Carbon is pulled from the atmosphere by plants, built into organic molecules, passed up the food chain through eating and being eaten, and eventually released back to the atmosphere through respiration and decomposition. The cycle is roughly balanced, or was, until we started burning fossil carbon that had been sitting outside the cycle for hundreds of millions of years.

Biofuels: ethanol, biodiesel, the tradeoffs

If plant-based organisms store solar energy in organic molecules and we can extract and burn them, then we have a renewable fuel. Sunlight hits plants. Plants make glucose or oils. We harvest those and combust them. We release carbon dioxide. Plants absorb that carbon dioxide and make more fuel. Carbon goes around in a circle, not upward in a straight line.

This is the promise of biofuels. And it is a genuine promise. But it comes with complications that are also written in chemistry and in land.

The most widely used biofuel is ethanol, a simple two-carbon molecule that you have met before in this book in the context of fermentation. When yeast breaks down glucose without oxygen, it produces ethanol and carbon dioxide. This is the same process that makes wine and beer, just scaled up enormously and applied to corn or sugarcane instead of grapes.

Corn-based ethanol is the dominant biofuel in the United States. Sugarcane ethanol is the dominant form in Brazil. The molecules are chemically identical. The difference is in the feedstock and the energy required to make it.

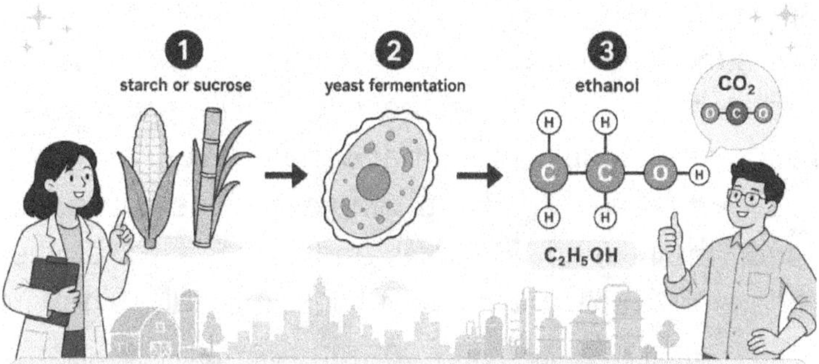

From sugar to ethanol -- fermentation pathway.

Diagram 9.8: From sugar to ethanol, fermentation pathway

Here is where the complication begins. Making ethanol from corn requires land that could be growing food. It requires water for irrigation. It requires nitrogen fertilizer, which is itself an energy-intensive product to manufacture. When you add up all those energy inputs, the net energy gain from corn ethanol is actually quite modest. Estimates vary, but most analyses put it at around 1.2 to 1.5 units of energy out for every unit of energy in. That is better than nothing, but it is not the rousing, positive energy that biofuel advocates initially hoped for.

Sugarcane ethanol does better. Because sugarcane grows more densely and contains more fermentable sugar per acre than corn, the energy return is closer to 8 units out for every unit in. This is why Brazil, with its tropical growing conditions perfectly suited to sugarcane, has had genuine success running its vehicle fleet on ethanol blends.

The other main biofuel is biodiesel, which is made not from sugars but from oils. Plant oils, animal fats, and used cooking oil can all be chemically processed to produce biodiesel, a fuel that behaves very similarly to petroleum diesel in diesel engines. Restaurants that collect their used fryer oil and send it to biodiesel processors are not being sentimental. They are participating in a genuine carbon recycling loop.

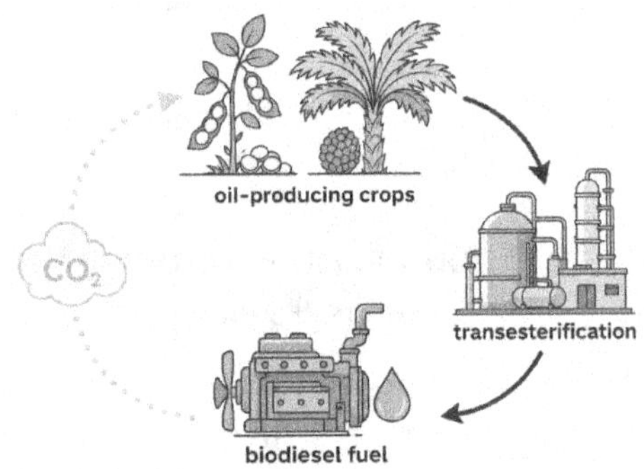

The biodiesel loop.

Diagram 9.9: The biodiesel loop

But biodiesel has its own tension. The most productive oil crops, soybeans in the United States, and palm oil in Southeast Asia, are land-hungry. When palm oil expansion requires clearing tropical forest, the carbon released from that forest soil and destroyed vegetation can exceed the carbon saved by displacing petroleum diesel by decades. The carbon math matters, and it depends enormously on what land is being used and how it was used before.

The honest summary of biofuels is this: they are real, they work, and they represent a meaningful part of the energy transition, especially where food and fuel crops can be separated, or where agricultural waste (corn stalks, rice husks, wood chips) is used as feedstock instead of the food crop itself. These second-generation, or cellulosic, biofuels are harder to process chemically but avoid the food-versus-fuel conflict entirely.

The dream version is algae. Algae grow in water, not on agricultural land. Some species are 50 percent oil by mass. They grow fast. They can be grown in brackish water or wastewater. In theory, algae biofuels could supply enormous amounts of energy without touching a single acre of food-growing land. In practice, scaling algae cultivation to fuel-supply levels has proven harder than optimists hoped in the 2000s. The chemistry is understood. The engineering is still catching up.

Hydrogen and the carbon-free future

Let's end with the molecule that has been called the fuel of the future for most of the past 50 years.

Hydrogen is not an energy source. This distinction matters enormously. Coal, petroleum, and natural gas are energy sources; they contain stored chemical energy, and we extract them from the ground. Hydrogen is an energy carrier. There is almost no free hydrogen on Earth. Every hydrogen atom you see is attached to something else: water, natural gas, or organic molecules. To get hydrogen fuel, you have to make it, and making it requires energy.

This is not a knockdown objection. Electricity is also an energy carrier. Nobody says electricity is not useful just because you have to generate it. The question is whether hydrogen is a convenient, efficient, and clean way to store and transport energy.

For many applications, the answer is genuinely promising.

A hydrogen fuel cell works by running the reverse of combustion. You feed hydrogen and oxygen into the cell. Inside the cell, a chemical reaction combines them to form water, thereby generating an electric current. No combustion. No carbon dioxide. The only exhaust is water vapor.

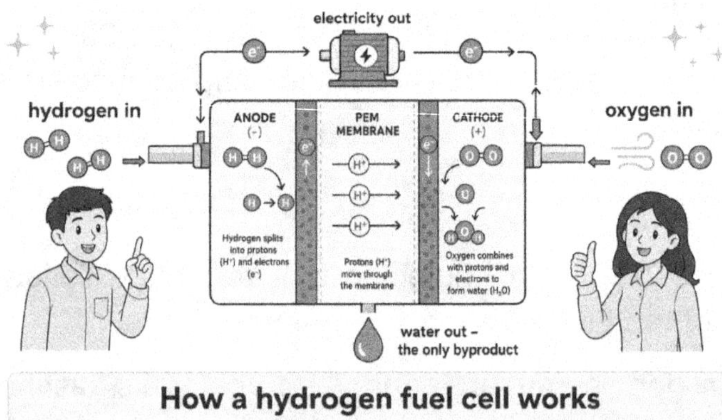

Diagram 9.10: How a hydrogen fuel cell works

The energy story is attractive. Hydrogen has the highest energy density per kilogram of any fuel, 120 megajoules per kilogram. A hydrogen-powered vehicle produces zero tailpipe carbon emissions. The only thing that comes out is water.

The storage problem is the villain of this story. Hydrogen gas is so low-density that you need either a very high-pressure tank or extreme cold to store enough of it to be useful as a vehicle fuel. High-pressure hydrogen tanks are heavy and must be engineered to extraordinary tolerances. Liquid hydrogen requires cryogenic storage and continuous energy input to maintain its temperature. Either option adds weight and cost that gasoline does not have.

The other issue is where the hydrogen comes from. Most hydrogen produced today is made by heating natural gas with steam, a process that releases significant carbon dioxide. This is called grey hydrogen, and it defeats much of the purpose. Green hydrogen, made by using renewable electricity to split water into hydrogen and oxygen, a process called electrolysis, is carbon-free, but currently expensive because renewable electricity is still more expensive than natural gas in most of the world.

The cost of green hydrogen is falling as renewable energy gets cheaper. Several countries, including Germany, Japan, Australia, and Chile, are making large bets on hydrogen as a major energy carrier for the second half of the 21st century. The places where hydrogen makes the most sense, long-haul trucking, shipping, steelmaking, and chemical production, are exactly the hard-to-electrify sectors where batteries are least practical.

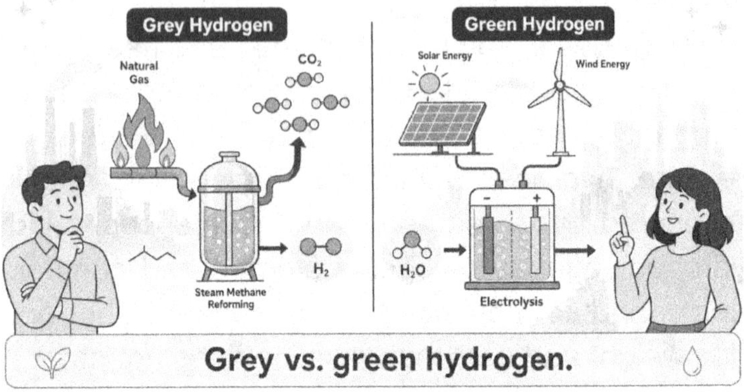

Grey vs. green hydrogen.

Diagram 9.11: Grey vs. green hydrogen

Here is the honest framing. The future of energy is almost certainly not one thing. It will be a portfolio: wind and solar electricity, stored in batteries where batteries are practical, stored in hydrogen where hydrogen is practical, supplemented by biofuels in sectors where liquid fuels are needed, and managed by a smarter grid than we have today. Every one of those solutions is a chemistry story. Every one of them involves carbon, either managing it, replacing it, or using it wisely.

The era of simply burning whatever we can dig up is coming to an end. The era of designed energy chemistry is opening. And you now understand the molecular logic behind every part of that transition.

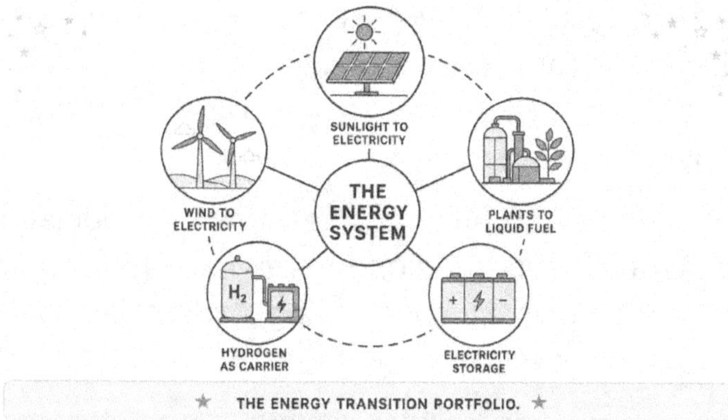

Diagram 9.12: The energy transition portfolio

Try this

Next time you eat something, think about where the energy came from. If you eat an apple, you are eating glucose that was assembled by a tree from carbon dioxide and rainwater using nothing but sunlight. The energy in that apple left the sun about eight minutes ago (as a photon), spent a few weeks being assembled into sugar by the tree, and is now about to be disassembled by your mitochondria into carbon dioxide and water again, with ATP collected along the way. You are running on solar power. You always have been.

Takeaways

- You can explain combustion at the molecular level: C-H and C-C bonds in hydrocarbons break, new bonds form in carbon dioxide and water, and the energy difference escapes as heat.

• You can explain why gasoline is genuinely hard to replace: its combination of high energy density per kilogram, liquid form, and easy storage is not matched by any current alternative.

• You can explain photosynthesis as the original solar cell: plants convert photon energy into chemical bond energy stored in glucose, using carbon dioxide and water as raw materials.

• You can explain cellular respiration as slow-motion combustion: your mitochondria break down glucose through hundreds of controlled enzymatic steps, capturing energy as ATP and releasing carbon dioxide and water, the same inputs and outputs as a flame, but at body temperature.

• You can articulate the core tradeoff of biofuels: they close the carbon loop in principle, but the land, water, and energy required to grow and process the feedstock determine whether the actual carbon math works out.

• You can explain why hydrogen fuel cells are promising and difficult: enormous energy per kilogram, zero carbon emissions at the point of use, but storage and production challenges that pure chemistry cannot yet overcome cheaply.

• You can describe the energy transition as a chemistry problem: the shift away from fossil fuels is not just about politics or engineering; it requires new molecular solutions for storage, conversion, and efficiency.

Looking ahead

Energy is one of the stories carbon tells. But there is another story, about what happens when we design molecules intentionally, rather than simply burning the ones we inherited. In Chapter 10, we turn to green chemistry: the idea that we can redesign how molecules are made, used, and discarded, building a molecular economy that produces less waste, uses less energy, and leaves the planet in better shape than we found it. The principles you have just learned, energy density, carbon cycles, and the logic of bonds, are exactly the tools you will need to understand why green chemistry is not just idealism. It is applied organic chemistry with a conscience.

10 The Next Chapter: Green Chemistry

You are scrolling through a news feed, the way you do, half-awake, coffee in hand. Three headlines catch your eye in quick succession. A new cancer drug that slips past healthy cells and destroys only tumors. A plastic bottle made from corn starch that composts in your backyard. A piece of software that solved a molecular puzzle scientists had been chasing for fifty years, and it did it over a long weekend. You skim past them, maybe pausing for a second. They feel like the kind of science news that belongs to a distant future, the kind of thing that happens in a lab somewhere you will never visit, understood by people who speak a language you never learned.

But you have learned that language. You have spent the length of this book building it, one carbon atom at a time. And those three headlines? They are not science fiction. They are organic chemistry, right now, today, doing what it has always done: rewriting the terms of what is possible for human life. The difference is that this time, the field is rewriting its own terms as well. The next chapter of organic chemistry is not just bigger and faster. It is wiser.

This is the chapter where we look forward, not with vague optimism, but with the specific, grounded optimism that comes from actually understanding what is happening at the molecular level. By the time you finish reading, you will be able to pick up a news story about a biodegradable plastic or a gene-editing therapy or an AI-designed drug and follow it, really follow it, from headline to molecule to consequence. That is not a small thing. That is the whole point.

The twelve principles of green chemistry

For most of its history, organic chemistry operated under a simple mandate: make the molecule. If the synthesis worked, if the yield was acceptable, if the product did what it was supposed to do, the job was done. What happened to the solvents, the byproducts, the waste streams, the energy? Those were someone else's problem. Someone in a yellow hazmat suit, probably.

That attitude built the modern world. It also left behind a considerable mess.

Green chemistry is the formal rethinking of that mandate. It asks not just "can we make it?" but "should we make it this way, and if not, what is a better way?" The field has a manifesto of sorts: twelve principles, first articulated by chemists Paul Anastas and John Warner in 1998, that describe what responsible molecular design looks like. We are not going to march through all twelve items in order, like items on a tax form. Instead, let's group them into the four big ideas that actually matter.

The first big idea: make less garbage.

Every reaction produces things you want and things you do not want. Traditional chemistry optimized for the first category and largely ignored the second. Green chemistry puts a name on the waste: atom economy. The concept is straightforward. If you start with ten atoms and you want eight of them in your product, what happens to the other two? They go somewhere. They become effluent, or hazardous waste, or a regulatory problem. A high-atom-economy reaction is one in which almost everything you put in ends up in the product you wanted to make. Nature is extraordinary at this. Enzymes in your liver conduct reactions with atom economies that would make a pharmaceutical chemist weep with envy.

The practical consequence is that green chemists design reactions backward from the product, asking at every step: What is this step generating that I do not need? Can a different reaction pathway get me to the same place with less waste? The best synthesis is neither the fastest nor the most elegant on paper. It is the one that produces the least stuff you did not ask for.

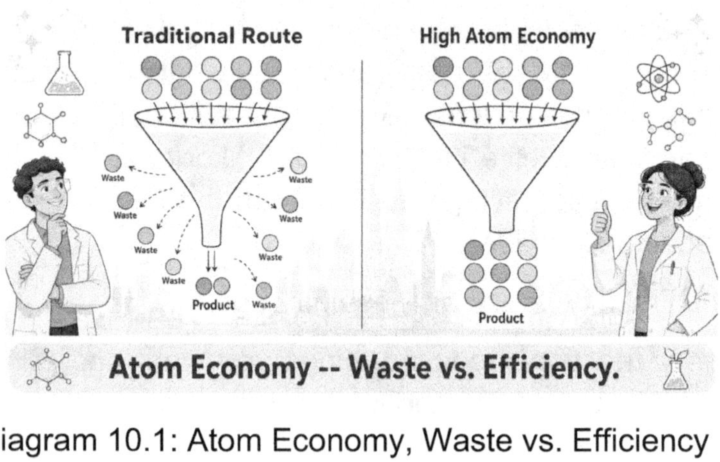

Diagram 10.1: Atom Economy, Waste vs. Efficiency

The second big idea: use safer ingredients.

The reagents and solvents that make reactions work are often the most dangerous things in the room. Many classic organic chemistry procedures run in solvents like dichloromethane or hexane, substances that are excellent at dissolving things and also excellent at dissolving your liver if you spend enough time around them. Green chemistry pushes toward safer solvents, which in practice often means water. Water is an extraordinary solvent for many biological reactions. It is cheap, abundant, and it will not give anyone cancer. Running reactions in water or in naturally derived solvents like ethanol or ethyl acetate is an active area of research, and the results have been surprisingly promising.

The same logic applies to the reagents themselves. If you can achieve a transformation with a mild, non-toxic catalyst rather than a vat of concentrated sulfuric acid, you should. Catalysts, by definition, are not consumed in reactions. They stick around to do the same job again. Using a catalyst is almost always greener than using a stoichiometric reagent, which is consumed in a one-to-one ratio with your starting material.

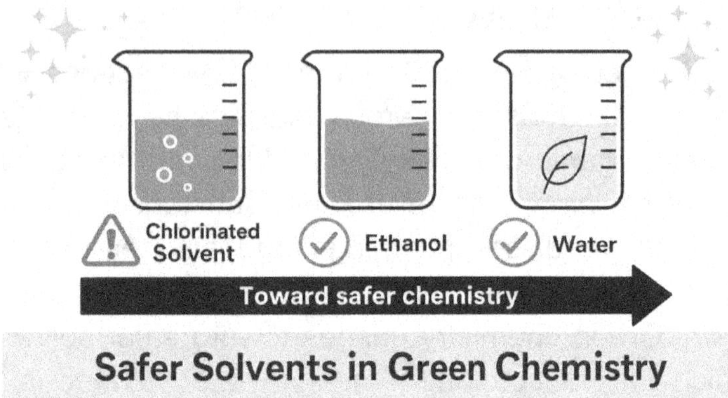

Diagram 10.2: Safer Solvents in Green Chemistry

The third big idea: use less energy.

Chemical reactions are often run at high temperature and high pressure because that is how you force molecules to do things they would rather not do. Green chemistry asks whether there is a gentler way. Enzymes, nature's catalysts, do their work at body temperature and atmospheric pressure. They manage this because they are extraordinarily good at lowering the activation energy barrier, the energy hill a reaction has to climb to proceed. Designing synthetic catalysts that can do the same thing, at room temperature, in water, using cheap starting materials, is one of the defining challenges and triumphs of modern green chemistry.

Photocatalysis is one exciting answer. Instead of heating a flask to drive a reaction, you shine light on it. The photocatalyst absorbs a photon and uses the energy to initiate the desired chemistry. The sun, it turns out, is a perfectly good energy source for a significant number of organic reactions. Chemists are only beginning to inventory which ones.

The fourth big idea: design for the end.

This is the one that most people miss. When you design a molecule, you are also designing its fate. A molecule that does its job and then persists in the environment for centuries, in a landfill, in a river, in the body fat of a polar bear, is a molecule that was only half-designed. Green chemistry demands that you consider degradation from the outset. Can the product be designed to break apart into harmless components once its useful life is over? For polymers, this is the question behind the entire biodegradable plastics industry. For pharmaceuticals, the question is what happens when a drug passes through the body and enters the water supply. The answers are not always simple. But now that chemists are asking the question at all, the answers are getting better.

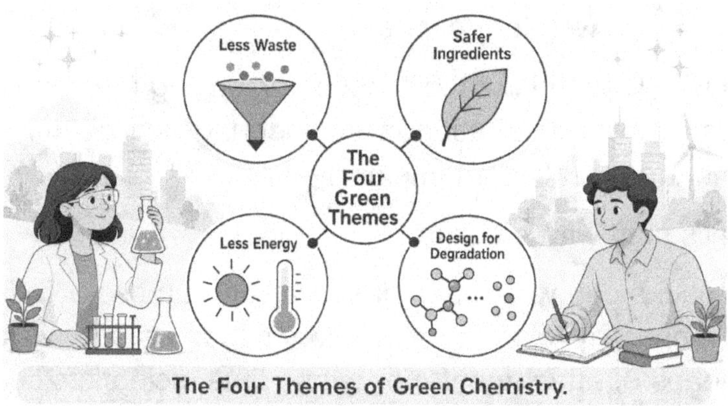

Diagram 10.3: The Four Themes of Green Chemistry

The twelve principles are a compass, not a checklist. No synthesis perfectly satisfies all of them. But they changed how chemists are trained to think, and that shift in thinking is already showing up in the products on store shelves and the drugs in clinical trials. The chemist of the next generation is not just someone who can make a molecule. They are someone who makes it responsibly.

Biomimicry: copying nature's playbook

Here is a fact that should recalibrate your sense of what is impressive. Life on Earth has been running chemistry experiments for approximately 3.8 billion years. That is not 3.8 billion years of random fumbling. That is 3.8 billion years of iterative optimization, with survival as the filter for quality. Every organism alive today is the heir to an unbroken chain of successful molecular designs. And when we look at those designs carefully, we find solutions to problems we have been struggling with for decades.

This is biomimicry: the practice of looking to biological systems for inspiration in engineering and chemistry, and then stealing shamelessly.

Let's start with a gecko.

A gecko can walk up a glass wall and across a ceiling without any glue, any suction cups, or any visible means of attachment. For a long time, scientists could not figure out how it did this. The answer, when it came, was delightful. The gecko's foot is covered in millions of tiny hair-like structures called setae, each of which branches into even tinier structures called spatulae. These spatulae are so small, and so numerous, and pressed so intimately against whatever surface the gecko is walking on, that van der Waals forces, the faint, transient attractive forces between any two molecules in proximity, add up to something significant. No chemistry required. Just geometry and proximity.

The lesson for materials science was immediate: Can we build a synthetic dry adhesive that works the same way? Researchers at several universities and companies have done exactly that, fabricating synthetic gecko-inspired materials from polymers with micro-structured surfaces. These materials stick to glass, release cleanly, and leave no residue. They have obvious applications in robotics, in medical bandages that do not tear skin, and in any situation where you want an adhesive that you can use over and over without it losing its grip. Nature solved this problem without a single molecule of glue. The synthetic version borrows the geometry and uses carbon-based polymers to replicate the nanostructure.

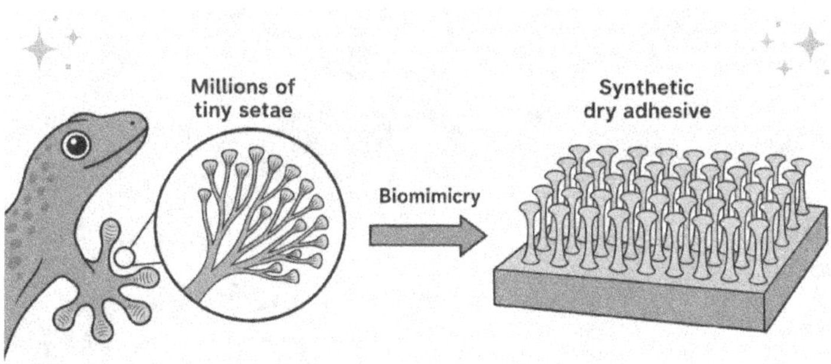

Gecko Foot to Dry Adhesive.

Diagram 10.4: Gecko Foot to Dry Adhesive

The lotus plant poses a different problem. Its leaves are famously self-cleaning: rainwater beads up into perfect spheres, rolls off, and carries any dirt or dust along. The surface of a lotus leaf is not smooth. It is covered in microscopic waxy bumps that minimize the contact area between the leaf and a water droplet. Because the droplet is barely touching the surface, it cannot spread out and stick. It rolls off, taking contaminants with it.

This observation gave rise to an entire category of superhydrophobic materials, surfaces engineered at the nanoscale to repel water the way a lotus leaf does. Spray-on coatings that make fabrics and glass self-cleaning are now commercially available. The structural principle was always in nature. The chemist's job was to see it, understand it, and translate it into a manufacturable material.

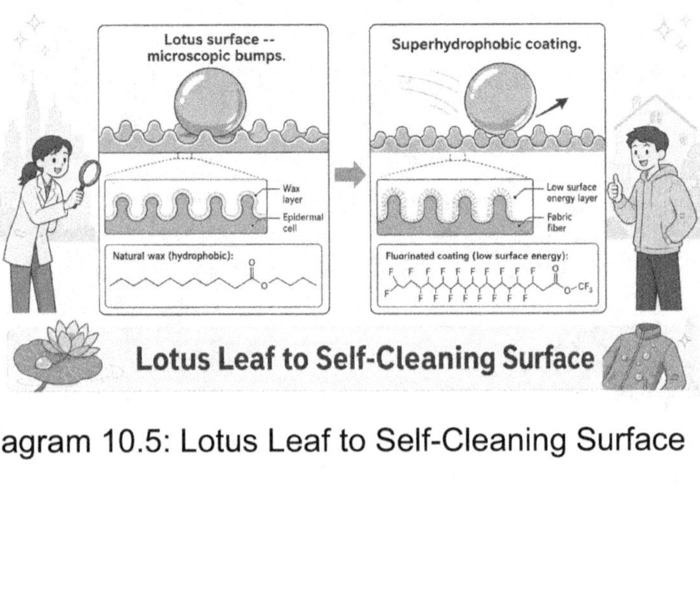

Diagram 10.5: Lotus Leaf to Self-Cleaning Surface

Spider silk deserves its own paragraph of astonishment. Weight for weight, it is stronger than steel and tougher than Kevlar. It stretches before it breaks, which is a combination of properties, simultaneously strong and elastic, that synthetic materials usually achieve only at great expense and complexity. A spider spins silk at room temperature, in water, from proteins. No extreme heat, no toxic solvents, no exotic catalysts. The silk fiber forms as the protein threads pack together in a specific molecular arrangement, with crystalline regions providing strength and amorphous regions providing elasticity.

For decades, the dream of synthetic spider silk seemed tantalizingly close and practically impossible. Spiders cannot be farmed the way silkworms can. They are territorial and cannibalistic, which makes them poor candidates for industrial production. The solution came from a combination of biotechnology and chemistry: engineer bacteria or yeast to produce the spider silk proteins, then spin those proteins into fibers using machines that mimic the mechanical conditions inside a spider's spinneret. The result is a material that inherits the mechanical properties of natural spider silk without requiring a single spider. It is entering the market now in surgical sutures, performance textiles, and body armor.

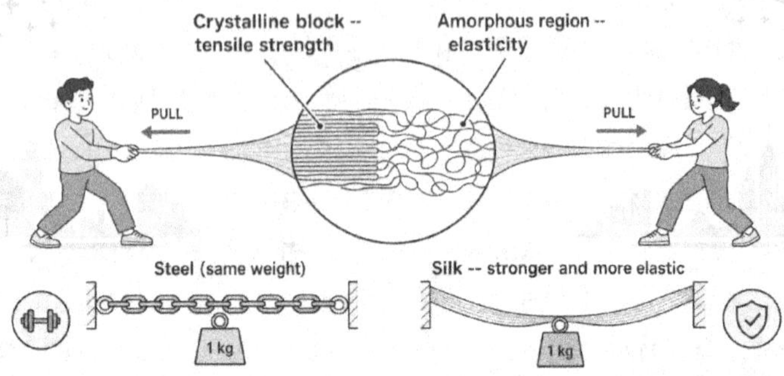

Spider Silk Structure and Properties.

Diagram 10.6: Spider Silk Structure and Properties

Then there is the mussel. If you have ever tried to pry a mussel off a rock at low tide, you know that they do not move willingly. They anchor themselves with a set of protein threads called byssus fibers, and those fibers hold fast in conditions of continuous saltwater, mechanical stress, and wave impact, conditions that destroy virtually every conventional adhesive. The key is a molecule called DOPA (a modified amino acid containing a catechol group), which forms extremely strong bonds with metal ions and mineral surfaces, even underwater.

Mussel-inspired adhesives based on DOPA chemistry are now being developed for medical applications: suturing internal wounds, sealing arterial tears, and bonding implants to bone. These are environments in which traditional surgical glue fails because the tissue is wet, mobile, and chemically complex. The mussel had that problem solved millions of years before anyone in a surgical suite did.

Nature's solutions are not always directly transferable. Sometimes they require extensive modification to work in a manufactured context. But the starting point, the insight that this problem is already solved, and here is where to look, is worth 3.8 billion years of competitive R&D that we did not have to pay for.

CRISPR and molecular editing

Let us be honest about something. When this book started, the idea of "molecular editing" would have sounded like magic. We are finishing it, and it still sounds a bit like magic, except that now you understand the chemistry well enough to see exactly why it works.

CRISPR-Cas9 is a molecular system borrowed from bacteria. Bacteria have immune memories: when a virus attacks them, they store a short sequence of that virus's DNA as a reference. If the same virus attacks again, the bacterium deploys a protein called Cas9, a molecular pair of scissors, guided by a short RNA molecule that matches the stored sequence. Cas9 finds the matching DNA sequence in the virus, cuts it, and inactivates the virus. It is a beautiful, simple system. And then scientists realized: if we write our own RNA guide, we can direct those scissors to cut any DNA sequence we choose. In any genome. Including a human one.

The chemistry here is the chemistry you already know. DNA is a polymer of nucleotides. The guide RNA is also a polymer, and it finds its target by the same base-pairing rules that govern all of molecular biology: A pairs with T (or U in RNA), G pairs with C. This is not exotic chemistry. It is the same hydrogen-bonding logic that determines how DNA replicates, how mRNA is transcribed, and how ribosomes read the genetic code. CRISPR borrowed that logic and weaponized it for precision targeting.

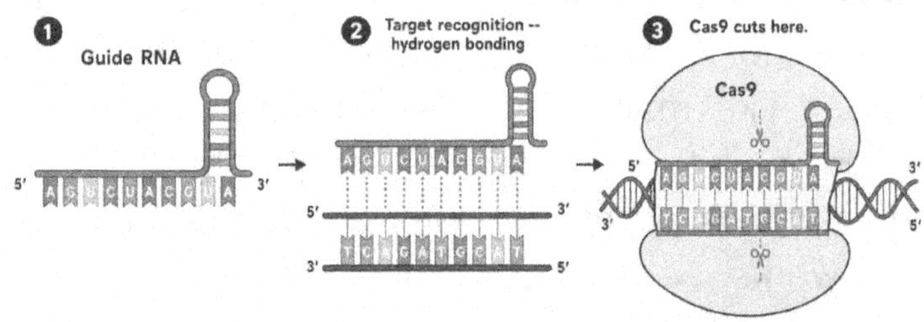

How CRISPR Finds Its Target

Diagram 10.7: How CRISPR Finds Its Target

The therapeutic applications are transforming medicine at a speed that can be hard to absorb. Sickle cell disease, caused by a single mutation in the hemoglobin gene, has been a life-shortening condition for which there has been no cure. In 2023, the first CRISPR-based therapy for sickle cell disease received regulatory approval. The therapy edits the patient's own bone marrow stem cells, correcting the faulty sequence at the source, managing the disease, suppressing the symptoms, and fixing the molecular sequence that was wrong.

The precision of CRISPR is what makes it both exciting and genuinely difficult. The guide RNA is specific to its target sequence, but no system is perfect. Off-target cuts (edits in the wrong place) are a real risk, and a significant portion of current research is devoted to making the system more accurate, more controllable, and more deliverable to specific tissues. Delivering the molecular machinery to exactly the right cells, in exactly the right amount, without triggering an immune response, is an engineering challenge that sits at the intersection of organic chemistry, polymer science, and cell biology.

The vector most commonly used to deliver CRISPR components into cells is the lipid nanoparticle, a tiny sphere of lipids (fat-like molecules you met earlier in this book) that can carry molecular cargo across cell membranes, which are themselves lipid bilayers. Like dissolves like, at the level of entire delivery vehicles. The same principle that governs which molecules can cross a membrane is being used to design the trucks that carry molecular editors inside cells.

Machine learning for drug discovery

The traditional path to a new drug looks something like this. A team of chemists starts with a biological target, a protein involved in a disease. Then it begins making molecules, one at a time, testing each one, modifying it based on what they learn, making another, testing again. This process takes roughly 12 years on average and, by most estimates, costs more than a billion dollars per drug that reaches the market. The odds that any given candidate molecule will survive clinical trials are in the single digits. It is a grind. It has also produced most of the medicines in your cabinet, so the grind has been worth it.

Machine learning is beginning to change the terms of that grind.

The premise is straightforward. The molecular properties of a compound, its shape, its charge distribution, its flexibility, and its tendency to bind to certain proteins can be described mathematically. Biological activity (does this molecule inhibit this enzyme? Does it block this receptor?) can be measured and recorded. If you have enough of those pairs, molecule descriptions, and biological results, you can train a model to find patterns in them that a human chemist might never notice. Then you can use that model to screen millions of virtual molecules in the time it would take to synthesize and test a few dozen real ones.

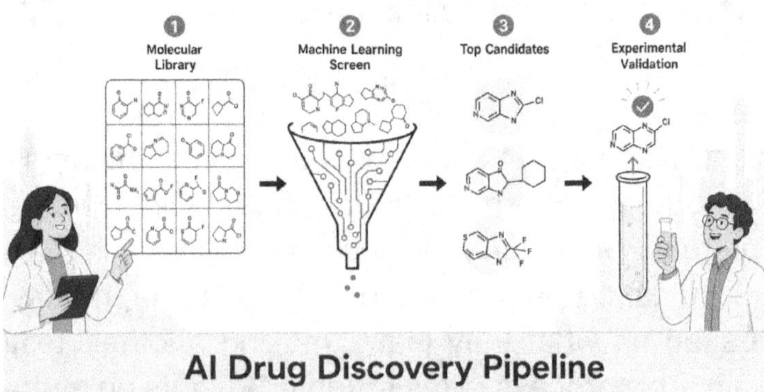

Diagram 10.8: AI Drug Discovery Pipeline

The approach is neither science fiction nor purely theoretical. Several drugs currently in clinical trials were identified by machine learning models that screened enormous virtual chemical spaces and surfaced molecules no human chemist had thought to try. The model did not understand chemistry the way a trained chemist does. But it found patterns in the data that pointed to unexplored corners of molecular space.

The most dramatic demonstration of what computation can do for molecular biology came from the protein folding challenge. Proteins are long chains of amino acids, and the function of a protein depends on the three-dimensional shape it folds into. Predicting that shape from the sequence alone had been one of the great unsolved problems in biology for fifty years. The reason it was hard: the number of possible configurations a protein chain can adopt is astronomical. For a modest-sized protein, the search space is larger than the number of atoms in the observable universe.

AlphaFold, a system developed by a team at DeepMind, cracked this problem in 2020. Given a protein's amino acid sequence, AlphaFold predicts its folded structure with accuracy that rivals experimental methods. Where experimental structure determination can take months or years and requires specialized equipment, AlphaFold produces a prediction in hours. The AlphaFold Protein Structure Database now contains predicted structures for hundreds of millions of proteins, many of which had never been structurally characterized in any laboratory on Earth.

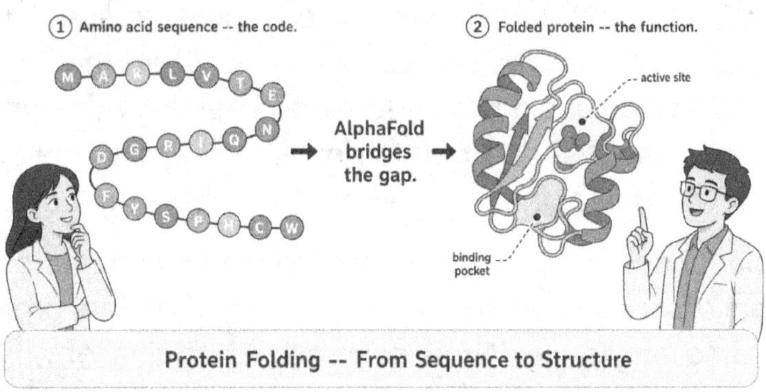

Diagram 10.9: Protein Folding, From Sequence to Structure

What does this mean for drug discovery? Everything. If you know the shape of a target protein, you can design molecules that fit into it the way a key fits a lock. Before AlphaFold, the shape of most proteins was unknown. Chemists were trying to design keys for locks they could not see. Now they can see the lock. The number of druggable targets in human biology has just expanded by an order of magnitude, as their structures are now known.

This is still in the early days. AlphaFold predicts the static structure of a protein, not how it moves or how it changes shape when a drug binds to it. The computational tools for those questions are improving, but they are not yet as transformative as the structure prediction itself. The drug still has to be made, tested, refined, and evaluated for safety in animals and humans. Human biology is still complicated. But the search space has shrunk from an ocean to a lake. That is a meaningful change.

Biodegradable materials by design

The plastics chapter of this book (you may remember it with a mixture of admiration and existential dread) told you how polymers transformed the twentieth century by being durable, cheap, and endlessly versatile. It also told you the price of that durability: plastic takes centuries to break down, and it is now everywhere, including inside us. The question for the twenty-first century is not whether to use polymers. It is how to design them so that their useful life and their environmental fate are matched.

This is polymer revolution 2.0.

The key insight is that biodegradability is not a property that polymers either have or lack by accident. It is a design parameter. You can build degradability into a polymer at the molecular level by choosing backbone linkages that biological enzymes can break.

Consider polylactic acid, or PLA. It is a plastic made from lactic acid, which is itself made by fermenting corn sugar or sugar beets. The polymer chain that forms from lactic acid contains ester linkages, the same type of chemical bond that enzymes in soil and water are already equipped to hydrolyze. Under the right temperature and humidity conditions, PLA breaks down into water and carbon dioxide within a matter of months. The carbon locked in the plastic came from plants that captured it from the atmosphere, and it returns to the atmosphere as the plastic degrades. The carbon cycle closes.

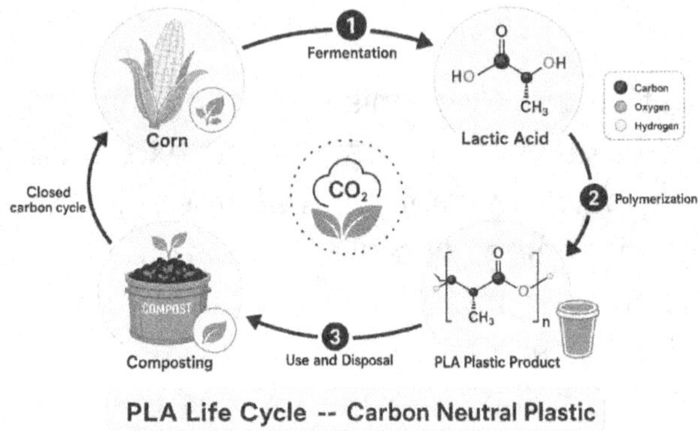

Diagram 10.10: PLA Life Cycle, Carbon Neutral Plastic

PLA has limitations. It requires industrial composting conditions (specific temperatures and humidity) to degrade quickly, and not every composting facility will accept it. It is not as strong or heat-resistant as conventional plastics. These are problems that current chemistry is actively solving through copolymerization (blending lactic acid with other monomers to tune mechanical properties) and through additives that accelerate breakdown in ambient conditions.

Polyhydroxyalkanoates, or PHAs, are another class of biodegradable polymers with a different origin story. Bacteria produce them naturally as energy-storage granules, the microbial equivalent of fat deposits. Under nutrient stress, certain bacteria accumulate PHAs within their cells in quantities that can exceed half their body weight. The polymers can be extracted and processed into plastic-like materials that biodegrade completely in marine environments, one of the most demanding tests for biodegradability, because the ocean is cold, dilute, and relentless. PHAs are already used in packaging, medical implants (they degrade harmlessly inside the body), and agricultural films.

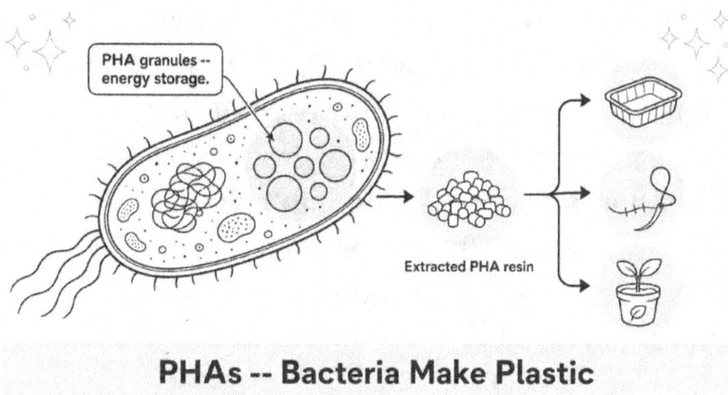

Diagram 10.11: PHAs, Bacteria Make Plastic

The deeper lesson is one of intention. Conventional plastics were designed to last and were only later confronted with the question of what happens when they are no longer needed. The new generation of materials is being designed with the end in mind, from the first molecular decision. That is green chemistry's fourth principle, design for degradation, made real and commercial.

There is genuine momentum here. The molecular tools, synthetic biology platforms, industrial composting infrastructure, and regulatory frameworks are all moving in the same direction. The polymer chemists of the next decade will routinely design both the material and its disposal pathway in the same sitting.

Your role: chemistry literacy is participation

We have covered a lot of ground. Twelve principles that are reshaping what responsible chemistry means. A billion years of biological R&D that chemists are beginning to appreciate and borrow from properly. A molecular scissors borrowed from bacteria that can edit the human genome. Algorithms that predict protein shapes in hours that took laboratories decades—plastics designed to disappear gracefully. The view from here is good.

But here is the thing. The most important decisions about these technologies will not be made in chemistry laboratories. They are going to be made in legislatures, in corporate boardrooms, in investment committees, in the offices of regulatory agencies, and at kitchen tables where people decide whether to buy the biodegradable packaging or the cheaper conventional one. Those decisions will be made by people who may or may not understand the molecular principles at stake. And the quality of those decisions will, in part, depend on how many people in the room actually understand what they are talking about.

You now understand what they are talking about.

That is not a small thing. Chemistry literacy is not a luxury for people who took advanced science classes.

It is a form of civic participation. When a city council debates whether to ban single-use plastics, you can evaluate the molecular arguments on both sides. When a pharmaceutical company announces a breakthrough in targeted drug delivery, you understand what the lipid nanoparticle is doing and why it matters. When a news story claims that a new biodegradable plastic is "carbon neutral," you know to ask whether that means it uses a closed-loop, biosourced monomer like PLA, or breaks down into smaller pieces faster.

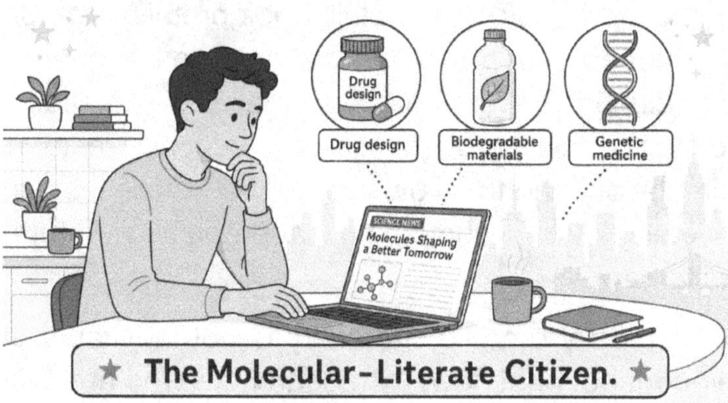

Diagram 10.12: The Molecular-Literate Citizen

The three headlines from the beginning of this chapter, the targeted cancer drug, the biodegradable bottle, and the AI that solved the protein mystery, are part of your world now. Not because you have become a chemist. But because you have learned to read the molecular level of reality, and once you can do that, you cannot unread it.

Consider what changes when you walk into a pharmacy. You see a shelf of ibuprofen bottles, and you do not just see a painkiller. You see a small molecule that fits into the active site of an enzyme that generates inflammatory signals, and you block it with the precision of a borrowed key. You pick up a bottle of sunscreen, and you see organic molecules with conjugated pi systems that absorb UV photons and release them as heat before they can damage your DNA. You see the chemistry that someone designed for you. And now you know enough to appreciate the design.

That is the transformation this book promised on page one. Not an expert. Not the ability to do the chemistry yourself. But genuine literacy: the ability to read the molecular world around you with curiosity instead of intimidation, and to follow the story of where that world is going with something approaching informed excitement.

The future of organic chemistry is not inevitable. It is being made, right now, by researchers who are asking better questions, by regulators who are setting smarter standards, by companies that are discovering that green is often good business, and by citizens who know enough to tell the difference. You are one of those citizens now.

The journey from a curious reader on page one to someone who can follow the molecular logic of a gene therapy or a biodegradable polymer is complete. The world is more legible than it was when you started. That is the gift that chemistry gives, not just to scientists, but to anyone who takes the time to look closely at the language written in carbon atoms. The Closing section of this book will draw together the full arc of that transformation, from the first carbon atom to this final horizon.

Try This

The next time you buy something in plastic packaging, look for the claim "biodegradable" or "compostable." Now ask the two questions this chapter equipped you to ask: What is the backbone? And under what conditions? A PLA cup labeled "compostable" will break down in an industrial composting facility at 55 degrees Celsius, but not in your kitchen compost or the ocean. A PHA-based material might biodegrade in marine conditions. The label is the beginning of the story. Now you have the molecular vocabulary to read past it.

Takeaways

• You can read a news story about green chemistry and assess whether a claim about biodegradability or atom economy is substantive or marketing, because you understand the molecular design principles behind each.

• You can explain why biomimicry is a serious scientific discipline, with specific examples: gecko foot adhesion uses van der Waals geometry, lotus leaf self-cleaning uses superhydrophobic surface architecture, and spider silk uses a two-phase crystalline and amorphous polymer structure.

• You can describe CRISPR-Cas9 in molecular terms: a guide RNA finds a target DNA sequence by base pairing, and the Cas9 protein cuts at that location, precision editing built on the same hydrogen-bonding logic that runs all of molecular biology.

• You can explain why AlphaFold was a scientific breakthrough: it solved the fifty-year protein folding problem, connecting amino acid sequence to three-dimensional structure, and opening vast new territory for drug discovery.

• You can distinguish between different classes of biodegradable polymers, PLA made from fermented plant sugars, PHAs made by bacteria, and understand that biodegradability is a molecular design choice, not an accident.

• You can articulate why chemistry literacy is a form of civic participation: the decisions shaping green chemistry, genetic medicine, and materials science will be made by non-scientists, and informed citizens make better choices.

• You can look at any headline about drug design, plastic pollution, gene therapy, or molecular technology and follow it from the news to the molecular level, because you have learned to read the language that the world is written in.

11 Closing

Think back to where you started.

You picked up this book as someone who had spent decades walking through a world assembled from molecules without being able to read a single one. The ingredient list on a shampoo bottle was a wall of foreign syllables. A drug recall made the news, and you understood the alarm but not the mechanism behind it. A friend who studied chemistry tried to explain something about chirality at a dinner party, and you smiled, nodded, and quietly changed the subject.

That was page one. You were a tourist, pressing your nose against the glass of a molecular world that seemed built for specialists only.

You are not on page one anymore.

Before and After at the Molecular Threshold.

Diagram C.1: Before and After at the Molecular Threshold

What You Walked Through

The journey had a shape. It was not random.

It started with carbon itself, that four-armed connector, agreeable and endlessly versatile, the one element that life auditioned for the starring role and then never let go. From there, you learned the language: the skeletal shorthand that chemists use to draw molecules the way musicians use staff notation, compact and portable and, once you know the rules, surprisingly readable.

Then came the personalities. Functional groups turned out to be exactly that: groups with distinct behaviors, predictable moods, and reliable tendencies. An alcohol acts like an alcohol. An ester smells fruity because it cannot help it. The carbon skeleton is the body; the functional group is the character. You can now tell them apart.

Chirality arrived next, and with it, one of the most quietly astonishing facts in all of chemistry: that two molecules can be made of identical atoms, bonded in an identical order, and still be fundamentally, consequentially different from each other, because one is the mirror image of the other, and biology is handed. You learned why thalidomide became a tragedy, and why modern drug developers spend years and billions of dollars making sure they have the right twin.

Reactions followed, told as the negotiations they actually are: molecules approaching each other, bonds breaking and forming, new arrangements emerging.

Then, drugs are designed at the atomic level to target a specific protein pocket in the body. Then polymers, spun into plastics and textiles, and DNA alike. Then, there is flavor and fragrance, the molecular poetry of every scent and taste you have ever loved. Then fuel, and the ancient sunlight locked inside carbon-hydrogen bonds, and the difficult question of what comes after. Then, finally, the frontier: green chemistry, biomimicry, and the accelerating science of designing better molecules on purpose.

That is a lot of ground covered. And you covered it without a single equation.

What You Can Do Now

Here is the part worth sitting with for a moment.

You can read a drug recall story in the news and understand, at least in outline, what went wrong. Was it a chirality problem, the wrong enantiomer slipping through? A functional group that behaves unexpectedly under the body's conditions? A reaction that the clinical trials did not catch at low doses? You may not know which one, but you know the categories. That is nothing. That is genuine comprehension.

You can look at a shampoo ingredient list and feel curiosity instead of intimidation. That silicone derivative? You know it is a polymer. That ester? You know it probably contributes to the scent. The fatty acid at the top of the list? You know it is doing the cleaning. The list has not changed. You have.

You can talk to a pharmacist, a doctor, a food scientist, or a chemistry teacher at something close to even footing. You will not bluff your way through the conversation. But you will not be lost in it either. You can ask better questions, understand more of the answers, and follow the thread of the explanation rather than waiting politely for it to end.

You can make choices about materials, foods, medications, and fuels with a new kind of clarity. Not perfect clarity. Organic chemistry is complicated, and the world is full of competing tradeoffs that no single book resolves. But you are choosing now with a map in your hand, not in the dark.

That shift, from alienation to ownership, was the promise this book made on the first page. It has kept it.

The Wider Map

This book is one waypoint.

The "Science for Curious Adults" series exists because the sciences are not separate kingdoms. They are different scales of the same wondrous physical world. "Quantum" descends to where carbon's four-bond capacity actually comes from, the strange, probability-governed behavior of electrons that makes chemistry possible in the first place. "Thermodynamics" zooms out to energy and entropy at every scale, giving you the mathematical backbone (in plain language) of why reactions go the direction they go. "Dark Matter" operates at the scale of galaxies, asking what most of the universe is actually made of, and

"Neuropharmacology" takes the receptor pharmacology you met in Chapter 6 and follows it deep into the brain.

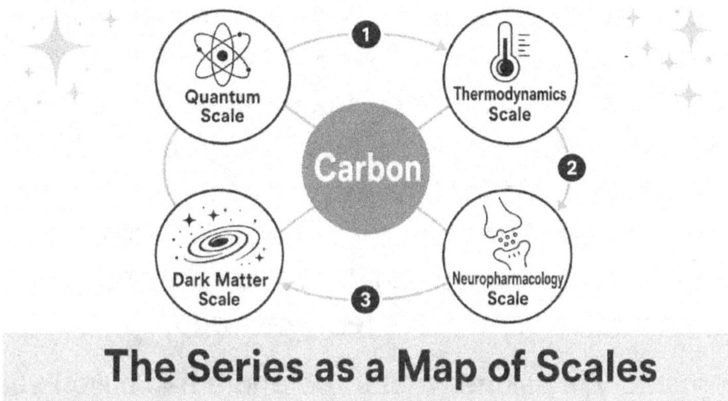

Diagram C.2: The Series as a Map of Scales

Each title is a different door into the same house. You can enter at any point, and the more titles you read, the more the connections between them light up. Readers who came to this book from "Quantum" will have found certain passages clicking into place with satisfying speed. Readers heading to "Neuropharmacology" next will find that the receptor pharmacology of Chapter 6 gives them a running start. The series is designed to reward exactly that kind of accumulation.

The Story Is Still Being Written

Organic chemistry is not a finished subject. It is a live one.

Green chemistry is redesigning industrial processes from the molecule up, asking at the design stage whether a compound can be made without toxic solvents, persistent byproducts, or reactions that require extreme conditions and enormous energy inputs. The principles are young. The progress is real.

AI-assisted drug discovery has compressed what used to take a decade of laboratory work into something closer to a long weekend of computation. AlphaFold's prediction of protein structures, a problem that had stumped the field for fifty years, happened in a matter of months once the right tool existed. Molecules are being designed now that would not have been imaginable as candidates ten years ago.

Biomimicry is asking what three and a half billion years of evolution have already figured out, and whether we can borrow the answer. Spider silk is stronger per unit weight than steel and is fully biodegradable. Lotus leaves are self-cleaning through a surface geometry so elegant that it has launched an entire field of materials research. The gecko's foot adheres without glue. Nature solved these polymer and surface problems long before we thought to ask the question.

You are now equipped to follow all of these stories as they unfold. You will encounter a headline about a new biodegradable plastic and know what the journalists mean by "polymer backbone." You will read about a new cancer drug targeting a specific enzyme and understand the molecular logic of the lock-and-key argument. You will hear "green chemistry" and know it is not a marketing phrase; it is a set of principles with molecular teeth.

The field is moving. You can move with it.

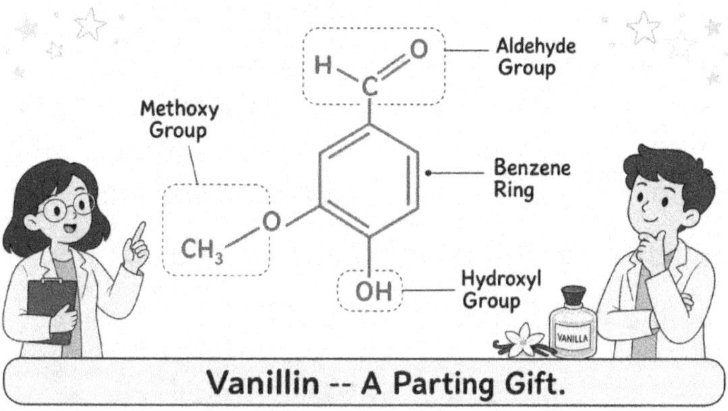

Diagram C.3: Vanillin, A Parting Gift

The World Is More Legible Now

There is a particular kind of pleasure that comes from understanding something you previously could not.

It is not a pleasure to show off. It is quieter than that. It is the pleasure of a world that has become more readable, where things that were opaque now have structure, where the labels on a bottle are no longer noise, where a news story about a molecule is a story you can actually follow.

That is what organic chemistry gives you when someone finally explains it well. Not a credential. Not a profession. A new way of seeing a world that was always this way, made of carbon, animated by chemistry, shaped by the behavior of molecules too small to see but too consequential to ignore.

The coffee you are drinking right now is a solution of roughly 800 volatile aroma compounds, a mild alkaloid (caffeine, a molecule you now know belongs to a family called xanthines), and a great deal of water, doing what polar molecules do best. The cup is a polymer. The table it sits on was once a tree, cellulose, nature's most abundant polymer, assembled from glucose monomers by a process powered by sunlight.

You did not know that at page one. You do now.

Enjoy the world. You can read it.

12 Glossary

This glossary is a quick-reference companion to the book, written in the same spirit as the chapters themselves. If a term stopped you mid-read, find it here. If you want to revisit an idea you enjoyed, welcome back.

A

Activation energy. The minimum amount of energy that molecules need before a reaction will start. Think of it as the height of the hill a ball must roll over before it can race downhill. Catalysts and enzymes are so useful precisely because they lower the energy barrier without changing what is on the other side.

Agonist. A molecule that binds to a receptor and switches it on, mimicking or amplifying the signal the body normally sends to that site. Many drugs are agonists, designed to produce a response that the body does not generate strongly enough on its own. See also: Antagonist, Receptor.

Alcohol. A family of molecules built around a carbon atom bonded to an OH (hydroxyl) group. Ethanol is the drinkable member of this family; methanol is its toxic cousin. The OH group makes alcohols polar, which is why they mix with water and often have recognizable smells. See also: Ethanol, Functional group.

Aldehyde. A functional group featuring a carbon atom double-bonded to oxygen and bonded to at least one hydrogen, always sitting at the end of a carbon chain. Aldehydes are responsible for many sharp, sweet, or green aromas. Vanillin, the molecule that gives vanilla its character, is an aldehyde. See also: Functional group, Vanillin.

Alkane. The simplest family of organic molecules: carbon and hydrogen atoms linked together by single bonds only, with no functional groups attached. Methane (natural gas), propane, and the long chains in gasoline are all alkanes. They are chemically quiet, which is both why they make great fuels and why they are so hard to break down in the environment.

Amine. A functional group built around a nitrogen atom bonded to the carbon skeleton. Amines tend to have strong, often fishy or sharp smells, and they are essential in biology because amino acids (the building blocks of proteins) all carry an amine group. Many drugs contain amine groups because they interact readily with receptors. See also: Functional group, Nitrogen.

Antagonist. A molecule that occupies a receptor without activating it, effectively blocking the body's own signals or other molecules from getting in. Antagonists are the "doorstop" drugs: they don't turn anything on, but they prevent the door from opening. See also: Agonist, Receptor.

Aromatic ring. A flat, ring-shaped arrangement of carbon atoms in which electrons are shared in an unusually stable way around the whole ring. Benzene is the classic example. The stability of aromatic rings makes them incredibly common in drugs, fragrances, and natural products. The "aromatic" name comes from the fact that many of these molecules do have strong smells, though not all do. See also: Functional group.

Aspirin. One of the world's oldest and most widely used drugs, originally derived from willow bark and later synthesized chemically. Aspirin works by permanently blocking an enzyme called COX, which is involved in producing the chemicals that cause pain and inflammation. It is a beautiful early example of how a well-understood small molecule can perform precise biological work.

Atom. The smallest unit of a chemical element that still behaves like that element. Each atom has a nucleus packed with protons and neutrons, orbited by electrons. Carbon's atom has six protons, which defines it as carbon. Everything in chemistry, and everything in this book, starts here.

Atom economy. A measure of how efficiently a chemical reaction uses its raw materials. A reaction with high atom economy converts most of what you put in into the desired product, generating little waste. Green chemistry prizes atom economy because chemical waste often results in hazardous byproducts or wasted energy. See also: Green chemistry.

Atomic number. The number of protons in the nucleus of an atom, and the number that defines which element you're looking at. Carbon's atomic number is 6, meaning every carbon atom has exactly six protons.

Change that number, and you're no longer talking about carbon. See also: Atom, Periodic table.

B

Backbone (carbon). The continuous chain of carbon atoms that forms the structural spine of an organic molecule. Everything else, the functional groups, the hydrogen atoms, the rings, attaches to this backbone. The length and shape of the backbone strongly influence how a molecule behaves. See also: Carbon, Functional group, Skeletal formula.

Bioavailability. How much of a drug actually reaches its target after you take it? A pill might contain 500 mg of a compound, but if only 100 mg survives digestion and absorption into the bloodstream, the bioavailability is 20 percent. Drug designers spend enormous effort tweaking molecular structure to improve this number. See also: Drug, Receptor.

Biodegradable. A material or molecule that living organisms (bacteria, fungi, enzymes) can break down into simpler substances over a reasonable timescale. Many conventional plastics are not biodegradable because their long, tightly packed carbon chains resist the biological machinery that would decompose them. PLA is one example of a plastic specifically designed to be biodegradable. See also: PLA, Polymer, Microplastic.

Biofuel. A fuel produced from biological material rather than from fossil deposits. Ethanol fermented from corn or sugarcane is the most familiar biofuel. At the molecular level, biofuels store energy in carbon-hydrogen bonds just as gasoline does. The difference is that the carbon they release when burned was recently pulled from the atmosphere, making the cycle potentially closer to carbon-neutral. See also: Combustion, Ethanol, Fermentation.

Biomimicry. The practice of designing new materials, molecules, or systems by copying solutions that nature evolved over billions of years. Spider silk, gecko adhesive, and the lotus leaf's self-cleaning surface are all biomimicry targets because nature solved those engineering problems long before humans tried. In chemistry, biomimicry often means building synthetic molecules that mimic what enzymes or proteins do naturally.

Bond (chemical). The force that holds two atoms together. Bonds form when atoms share or transfer electrons in a way that gives both parties a more stable arrangement. Without bonds, there are no molecules. With bonds, there is everything.

Bond (covalent). A chemical bond in which two atoms share a pair of electrons. This is the dominant bond type in organic chemistry. Carbon forms four covalent bonds, and the variety of ways it can arrange them is the root of organic chemistry's extraordinary richness. See also: Bond (single, double, triple), Carbon.

Bond (single, double, triple). Single bonds share one pair of electrons and allow rotation around the bond axis. Double bonds share two pairs of electrons, which makes them stronger and locks nearby atoms into a flat arrangement. Triple bonds share three pairs and are the strongest of the three, appearing in molecules like acetylene—the more bonds between two carbons, the shorter and stronger the connection.

C

Carbon. The element with six protons in its nucleus has a uniquely social personality. Because carbon can form four stable bonds at once, it builds the longest and most varied chains of any element on the periodic table. Life on Earth chose carbon for exactly this reason, and the entire discipline of organic chemistry exists to explore what those four bonds can do. See also: Atomic number, Bond (covalent), Organic chemistry.

Carbon cycle. The global circulation of carbon atoms through the atmosphere, living organisms, soil, and oceans. Plants pull carbon dioxide out of the air during photosynthesis. Animals eat plants and exhale carbon dioxide. When organisms die, their carbon returns to the soil. Fossil fuels are ancient carbon that was removed from this cycle millions of years ago. Burning them injects it back very quickly. See also: Combustion, Photosynthesis, and Cellular Respiration.

Carbon-12. The most abundant form of carbon found in nature has six protons and six neutrons in its nucleus. It is the stable, everyday version of carbon that makes up your bones, your dinner, and the graphite in a pencil. Scientists use carbon-12 as the reference standard for measuring atomic masses.

Carboxylic acid. A functional group that combines a carbon double-bonded to oxygen and also bonded to an OH group. Carboxylic acids are responsible for the sharp smell of vinegar (acetic acid) and the tang of citrus fruits (citric acid). They are also critical in biochemistry because fatty acids and amino acids both carry this group. See also: Functional group, Fatty acid.

Catalyst. A substance that speeds up a chemical reaction without being consumed in the process. Catalysts work by providing an easier pathway between the starting materials and the products, thereby lowering the activation energy required. Enzymes are nature's catalysts. Industrial chemistry depends on synthetic catalysts to manufacture everything from plastics to fertilizers. See also: Activation energy, Enzyme.

Cellulose. The most abundant organic polymer on Earth, making up the rigid cell walls of plants and the structural fiber in wood, cotton, and paper. Cellulose is a long chain of glucose molecules, but humans lack the enzyme needed to break it apart, which is why we can't digest wood. Termites and certain gut bacteria can. See also: Glucose, Polymer.

Cellular respiration. The process by which living cells extract energy from glucose by breaking it down, step by step, in the presence of oxygen. The products are carbon dioxide, water, and a great deal of usable energy in the form of ATP. It is essentially a very slow, controlled version of combustion, the same bonds being broken, but done gently enough that the cell can capture the energy rather than just releasing it as heat. See also: Combustion, Glucose, Photosynthesis.

Chemical reaction. An event in which chemical bonds break, reform, or both, converting starting materials (reactants) into new substances (products). Reactions occur because the products are more stable or because energy is added to drive the change. Every drug effect, every metabolic process, every moment of digestion is a chemical reaction. See also: Reactant, Activation energy, Catalyst.

Chiral center. A carbon atom in a molecule that is bonded to four different groups, making it the point at which molecular handedness originates. When a carbon has four distinct attachments, there are two ways to arrange them in space, and those two arrangements are mirror images of each other. Thalidomide's tragic history traces directly to a single chiral center in that molecule. See also: Chirality, Enantiomer, Thalidomide.

Chirality. The property of a molecule that cannot be superimposed on its mirror image. Like your hands, which look identical but cannot fit into the same glove, chiral molecules come in left- and right-handed versions. In biology, this matters enormously because receptors and enzymes have their own handedness, so only one version of a chiral drug may fit. See also: Chiral center, Enantiomer, Stereochemistry.

Combustion. The reaction between an organic molecule and oxygen that releases energy as heat and light, producing carbon dioxide and water. It's what happens when you burn wood, light a candle, or drive a car. At the molecular level, combustion breaks the carbon-hydrogen and carbon-carbon bonds in fuel and forms the much stronger bonds in CO_2 and water, releasing the energy difference as heat. See also: Carbon cycle, Cellular respiration, Energy density.

CRISPR. A molecular editing tool, borrowed from bacterial immune systems, that allows scientists to cut and modify DNA at precise locations. CRISPR uses a guide molecule to find a specific sequence in the genome and a protein (Cas9) to cut it. From a chemistry standpoint, it is a marvel of molecular precision: a biological pair of scissors that can find one specific word in a library of billions of pages.

D

Diamond. One of the allotropes of carbon, in which each carbon atom is bonded to four others in a rigid three-dimensional lattice. This arrangement makes diamond the hardest naturally occurring substance on Earth. It is the same element as the graphite in a pencil lead. Same atom. Completely different personality, shaped entirely by how the bonds are arranged. See also: Carbon, Graphite, Graphene.

DNA. Deoxyribonucleic acid. The long polymer molecule that stores genetic information in every living cell is wound into the famous double helix. DNA is built from four monomer units (adenine, thymine, guanine, cytosine) arranged in sequences that encode the instructions for building and running an organism. From a chemistry standpoint, DNA is nature's most impressive demonstration of what a polymer can do. See also: Monomer, Polymer, Protein.

Drug. In the context of this book, a molecule is deliberately designed or selected to interact with a specific biological target and produce a therapeutic effect. The molecular size, shape, functional groups, and chirality of a drug all determine whether it reaches its target, binds to it correctly, and performs as intended without causing harm. See also: Agonist, Antagonist, Bioavailability, Lead compound, Receptor.

E

Electron. A negatively charged particle that orbits the nucleus of an atom. Electrons are the currency of chemistry: chemical bonds form when atoms share or transfer electrons, and nearly every chemical event is, at its core, a rearrangement of electrons. Carbon's six electrons, particularly the four in its outer shell, are what make its four-bond personality possible.

Element. A pure substance made of only one type of atom. There are 118 known elements, each defined by its atomic number (number of protons). Carbon is element 6. Hydrogen is element 1. All organic molecules are built from a small subset of the periodic table, mostly carbon, hydrogen, oxygen, and nitrogen. See also: Atom, Periodic table.

Elimination reaction. A type of reaction in which atoms or groups are removed from a molecule, typically creating a new double bond in the process. Think of it as the molecule shedding some of its cargo to become lighter and more reactive. Elimination reactions are one way chemists introduce double bonds that give molecules new properties and reactivities. See also: Chemical reaction, Substitution reaction.

Enantiomer. One of a pair of mirror-image molecules that are identical in every way except for the arrangement of atoms around a chiral center. Enantiomers have the same atoms, bonds, and connectivity, but they rotate polarized light in opposite directions and can have dramatically different biological effects. Spearmint and caraway owe their distinct smells to a pair of enantiomers called carvone. See also: Chiral center, Chirality, Racemic mixture, Thalidomide.

Energy density. The amount of energy stored in a given volume or mass of fuel. Gasoline has a very high energy density, which is why it has been so hard to replace in vehicles. Hydrogen has an even higher energy density by mass, but is difficult to store compactly. Understanding energy density at the molecular level means understanding which bonds store the most energy when they break. See also: Biofuel, Combustion.

Enzyme. A protein that acts as a biological catalyst, speeding up specific chemical reactions inside living cells. Enzymes are exquisitely selective, typically catalyzing only one reaction or one family of reactions. Without enzymes, the chemical reactions that sustain life would proceed far too slowly to support it. Many drugs work by specifically blocking an enzyme. See also: Catalyst, Protein, Chemical reaction.

Ester. A functional group is formed when a carboxylic acid and an alcohol react, releasing a water molecule. Esters are responsible for many of the fruity smells and flavors in nature. The smell of bananas comes from isoamyl acetate, an ester. Aspirin itself is an ester. They are among the most pleasing-smelling functional groups in the organic world. See also: Alcohol, Carboxylic acid, Functional group.

Ether. A functional group in which an oxygen atom is bonded to two carbon chains. Ethers are relatively unreactive and readily dissolve many organic compounds, which is why they were once widely used as anesthetics and continue to serve as solvents in chemical synthesis. Familiar ether smells are sweet and slightly sharp. See also: Functional group, Solvent.

Ethanol. The specific alcohol found in alcoholic beverages has two carbon atoms and a single OH group. Ethanol is also used as a biofuel and as a solvent. At the molecular level, its small size and OH group allow it to mix freely with water, which is why it dilutes so readily and why the body can process it (slowly). See also: Alcohol, Biofuel, Fermentation, Methanol.

F

Fatty acid. A long-chain carboxylic acid, typically with 12 to 20 carbon atoms in its backbone. Fatty acids are the molecular building blocks of fats and oils. Saturated fatty acids have no double bonds; unsaturated fatty acids have one or more. This structural difference is what separates a solid butter from a liquid olive oil at room temperature. See also: Carboxylic acid, Saturated vs. unsaturated, Triglyceride.

Fermentation. The process by which yeast and certain bacteria convert sugars into ethanol and carbon dioxide in the absence of oxygen. Fermentation is one of humanity's oldest chemical technologies. At the molecular level, yeast uses a sequence of enzyme-driven reactions to extract energy from glucose, releasing ethanol as a byproduct. Wine, beer, and bread all depend on this process. See also: Enzyme, Ethanol, Glucose, Yeast.

Functional group. A specific arrangement of atoms within a molecule that determines how that molecule behaves. Functional groups are the "personality" of an organic molecule. An alcohol group makes a molecule polar and water-soluble. A carboxylic acid group makes it acidic and sharp-smelling. A molecule's carbon backbone is its body; its functional groups are its character. See also: Alcohol, Aldehyde, Amine, Carboxylic acid, Ester, Ether, Ketone.

G

Glucose. A six-carbon sugar is the primary fuel molecule for most living things. Photosynthesis builds glucose from carbon dioxide and water. Cellular respiration tears it back down to extract the stored energy. Glucose is also the monomer from which the natural polymers starch and cellulose are built. See also: Cellulose, Cellular respiration, Fermentation, Photosynthesis, Sugar.

Graphene. A single, one-atom-thick layer of carbon atoms arranged in a flat hexagonal lattice, like a sheet of chicken wire made of carbon. Graphene is the strongest material ever measured, an excellent conductor of electricity, and almost completely transparent. It is essentially one layer peeled off of graphite. What a difference a dimension makes. See also: Carbon, Diamond, Graphite.

Graphite. An allotrope of carbon in which atoms are arranged in flat sheets that stack loosely on top of one another. The weak forces between layers mean the sheets slide easily, which is why graphite feels slippery and why it marks paper as pencil "lead." Graphene is one of those sheets, isolated. See also: Carbon, Diamond, Graphene.

Green chemistry. A philosophy and set of design principles for making chemistry cleaner, safer, and less wasteful. Green chemistry asks chemists to use renewable starting materials, avoid hazardous solvents, design molecules that degrade after use, and minimize waste at every stage. It is not about sacrificing performance. It is about designing better chemistry from the start. See also: Atom economy, Biodegradable, Solvent.

H

Heteroatom. Any atom in an organic molecule that is not carbon or hydrogen. Oxygen, nitrogen, sulfur, and chlorine are the most common heteroatoms. They are called "hetero" because they are different from the carbon-hydrogen backbone. Functional groups are almost always built around heteroatoms, and they are the atoms that give organic molecules most of their chemical personality. See also: Functional group.

Hydrocarbon. A molecule made entirely of carbon and hydrogen, with no heteroatoms. Methane, ethane, benzene, and the molecules in gasoline are all hydrocarbons. They tend to be nonpolar, meaning they don't mix with water and are flammable. When you burn a hydrocarbon completely, the products are carbon dioxide and water. See also: Alkane, Combustion, Heteroatom.

Hydrogen. The lightest and most abundant element in the universe, with just one proton and usually one electron. In organic chemistry, hydrogen atoms are almost everywhere, bonded to carbon throughout the skeleton. In hydrogen bonds, hydrogen acts as a tiny bridge between electronegative atoms. Molecular hydrogen (H2) is being explored as a clean energy carrier. See also: Hydrogen bond.

Hydrogen bond. A relatively weak but very important attraction between a hydrogen atom bonded to a highly electronegative atom (typically oxygen or nitrogen) and another nearby electronegative atom. Hydrogen bonds are what give water its unusually high boiling point, what holds DNA's double helix together, and what dictates how proteins fold. They are not true chemical bonds, but their collective influence is enormous. See also: Hydrogen, DNA, Protein.

I

Ibuprofen. A widely used pain reliever and anti-inflammatory drug, related structurally to aspirin but developed independently. Ibuprofen works by blocking the same COX enzymes as aspirin, but it does so reversibly. Its molecule is chiral, though the drug sold in pharmacies is typically a 50-50 mixture of both enantiomers. See also: Aspirin, Chiral center, Drug, Enantiomer.

Inorganic chemistry. The branch of chemistry concerned with molecules and compounds that are not based on carbon-carbon or carbon-hydrogen bonds. Metals, minerals, salts, and most of the geologically sourced materials in the world fall under inorganic chemistry. The boundary with organic chemistry is not perfectly sharp, but the distinction remains a useful map. See also: Organic chemistry.

Ion. An atom or molecule that has gained or lost one or more electrons, giving it a net electrical charge. Positively charged ions (cations) have lost electrons; negatively charged ions (anions) have gained them. Table salt, for example, is composed of positively charged sodium ions and negatively charged chloride ions. Ions are critical in biological systems, especially in nerve signaling and membrane transport.

IUPAC. The International Union of Pure and Applied Chemistry is the body responsible for standardizing the naming of chemical compounds worldwide. IUPAC names are systematic, unambiguous, and often terrifyingly long, which is why trivial names like "aspirin" or "ibuprofen" persist in everyday use. When you see a tongue-twisting chemical name on a label, IUPAC is behind it.

K

Ketone. A functional group in which a carbon atom is double-bonded to oxygen and sits in the middle of a carbon chain (rather than at the end, as in an aldehyde). Acetone, the solvent in nail polish remover, is one of the simplest ketones. Its sweet, sharp smell is characteristic of the ketone group. See also: Aldehyde, Functional group.

L

Lead compound. In drug development, the first molecule that shows genuine promise against a biological target becomes the starting point for optimization. Finding a lead compound is one of the most exciting moments in drug discovery. It may come from screening thousands of molecules, from a natural product, or from rational molecular design. Everything after that is chemistry: adjusting the structure to improve potency, safety, and bioavailability. See also: Drug, Natural product.

LEGO connector analogy. The image used in this book explains carbon's four-bond capacity. Imagine carbon as a LEGO brick with four connection points, each able to snap to another atom's connector. Because carbon can link to other carbons this way, it can build chains, branches, and rings of any length. No other element has this combination of four bonds, moderate bond strength, and the ability to bond with itself so readily.

M

Maillard reaction. The cascade of chemical reactions that occurs between amino acids and sugars when food is heated, producing hundreds of new flavor and aroma compounds, and the characteristic brown color of seared meat, toasted bread, and roasted coffee. Named for French chemist Louis-Camille Maillard, who first described it in 1912. The Maillard reaction is not a single reaction but many hundreds of reactions happening simultaneously, which is why browned food tastes so much more complex than boiled food. See also: Amino acid (see Amine), Sugar.

Mass (molecular). The total mass of a molecule is calculated by adding up the masses of all its constituent atoms. Molecular mass is usually expressed in daltons or atomic mass units. Small drug molecules typically weigh somewhere between 150 and 600 daltons. Polymers can have molecular masses in the millions. Size matters in chemistry because it affects how a molecule moves, dissolves, and interacts with biological machinery.

Methanol. The simplest alcohol: one carbon, one OH group. Unlike ethanol, methanol is highly toxic to humans even in small amounts. The reason is metabolic: the body converts methanol into formaldehyde and formic acid, which destroy the optic nerve and cause liver failure. This cautionary chemical story illustrates how profoundly two molecules can differ when their carbon skeletons differ by just a single atom. See also: Alcohol, Ethanol.

Microplastic. A plastic particle smaller than 5 millimeters that forms when larger plastic objects break into fragments under UV radiation, physical abrasion, or other environmental stresses. Because the underlying polymer molecules do not degrade, microplastics accumulate everywhere, from the deepest ocean trench to human lung tissue. Their chemical stability, the very property that made plastics useful, is what makes microplastics so persistent. See also: Biodegradable, Polymer, Polyethylene.

Molecule. Two or more atoms held together by chemical bonds, behaving as a single unit. Water is a molecule. So is aspirin, DNA, and a strand of nylon. Molecules can be as small as two atoms or as large as millions, and their properties emerge not just from which atoms they contain but from how those atoms are arranged. See also: Atom, Bond (covalent).

Monomer. The small, repeating molecular unit that chains together to build a polymer. Ethylene is the monomer of polyethylene. Glucose is the monomer of starch and cellulose. Amino acids are the monomers of proteins. The properties of a polymer depend not only on the properties of its monomer but also on how many units are strung together and how they arrange in space. See also: Polymer, DNA, Protein.

N

Natural product. A molecule produced by a living organism, plant, fungus, bacterium, or animal, that often has significant biological activity. Penicillin was originally a natural product from a mold. Aspirin traces its origins to salicylic acid from willow bark. Natural products have been the starting point for a remarkable fraction of all pharmaceuticals. See also: Lead compound, Penicillin.

Nitrogen. Element 7 is the fourth most common element in organic molecules after carbon, hydrogen, and oxygen. Nitrogen atoms carry lone pairs of electrons that make them particularly reactive and important in biological chemistry. Amines, amides, and the base pairs of DNA all depend on nitrogen. The nitrogen in your proteins and DNA ultimately comes from the atmosphere via bacteria that can "fix" it into usable compounds. See also: Amine, DNA, Heteroatom.

Nylon. One of the first fully synthetic polymers was developed at DuPont in the 1930s. Nylon is a polyamide, meaning amide bonds link its monomer units, the same type of bond that links amino acids in proteins. It was designed to mimic silk, and it succeeded remarkably. Nylon's development marked the moment humanity proved that useful natural materials could be replicated and eventually surpassed by deliberate molecular design. See also: Monomer, Polymer, Protein.

O

Olfactory receptor. A protein in the lining of the nose that binds specific airborne molecules and triggers a nerve signal that the brain interprets as a smell. Humans have roughly 400 types of olfactory receptors, and fragrances work by activating combinations of these receptors in characteristic patterns. The reason one molecule smells like banana and another like gasoline is that they fit different receptor combinations. See also: Receptor, Terpene.

Organic chemistry. The branch of chemistry devoted to molecules built on a carbon skeleton. "Organic" originally meant "derived from living things," but the term now includes any carbon-based molecule, whether from a tree, a laboratory, or a petroleum refinery. Organic chemistry is the science underlying medicine, plastics, food, fuel, and life itself. See also: Carbon, Inorganic chemistry.

Oxygen. Element 8, the most electronegative common element in organic chemistry, is one of the most important heteroatoms. Oxygen appears in nearly every major functional group: alcohols, carboxylic acids, esters, ethers, aldehydes, and ketones all contain oxygen. Its strong electron-withdrawing effect is what makes these functional groups reactive, polar, and biologically significant. See also: Heteroatom, Functional group.

P

Penicillin. A family of antibiotic compounds discovered by Alexander Fleming in 1928, originally produced by the mold Penicillium notatum. Penicillin works by turning off an enzyme that bacteria use to build their cell walls. Without the enzyme, new bacteria cannot form proper cell walls and burst. It is a masterclass in how a small molecule, correctly shaped, can find and turn off a specific biological target. See also: Drug, Enzyme, Natural product.

Periodic table. The organized chart of all known chemical elements, arranged by atomic number and grouped according to similar chemical properties. In this book, the periodic table is treated not as a memorization exercise but as a personality chart: elements in the same column tend to form the same number of bonds and have similar chemical behaviors. Carbon's position in the table tells you immediately that it forms four bonds. See also: Atom, Atomic number, Element.

Phenol. A molecule in which an OH (hydroxyl) group is attached directly to an aromatic ring. Phenols are more acidic than ordinary alcohols because the ring stabilizes the resulting negative charge. Many naturally occurring compounds in plants are phenols, including the active ingredient in aspirin's ancestor, salicylic acid. See also: Alcohol, Aromatic ring.

Photosynthesis. The process by which plants, algae, and some bacteria use sunlight to convert carbon dioxide and water into glucose and oxygen. Photosynthesis is the original solar energy capture system, building energy-rich carbon-carbon bonds using nothing but light. It is the foundation of virtually all food chains on Earth and the source of the oxygen in the atmosphere. See also: Carbon cycle, Cellular respiration, Glucose.

PLA (polylactic acid). A biodegradable polymer made from lactic acid, which is itself derived from fermented plant sugars such as corn starch. PLA is used in compostable cups, packaging, and medical sutures.

Unlike conventional plastics, its ester bonds can be broken apart by microorganisms under the right conditions. It is one of the most successful examples of designing biodegradability into a polymer from the molecular level up. See also: Biodegradable, Ester, Polymer.

Polymer. A large molecule built by linking many smaller repeating units (monomers) together in a chain. Polymers can be natural (DNA, proteins, cellulose, rubber) or synthetic (polyethylene, nylon, PVC). Their properties, strength, flexibility, solubility, and stability depend on the identity of the monomer, the length of the chain, and how the chains pack together. See also: Monomer, Polyethylene, DNA, Protein.

Polyethylene. The world's most produced plastic is made by linking thousands of ethylene molecules into long carbon chains. Low-density polyethylene (LDPE) is flexible and used in plastic bags. High-density polyethylene (HDPE) is stiffer and used in bottles and pipes. The difference in properties comes not from different atoms but from how the chains pack together. See also: Monomer, Polymer.

Polypropylene. A plastic similar to polyethylene but with a methyl group attached to every other carbon in the backbone. This small structural difference makes polypropylene stiff and heat-resistant enough to withstand dishwasher cycles, which is why it is used in food containers marked with the recycling symbol 5. See also: Polymer, Polyethylene.

Polystyrene. A polymer made from styrene monomers, each carrying an aromatic ring as a side group. Those bulky rings prevent the chains from packing tightly, giving polystyrene a glassy, rigid character. Expanded polystyrene foam (the material of coffee cups and packaging peanuts) is polystyrene with air bubbles trapped inside. See also: Aromatic ring, Monomer, Polymer.

Protein. A polymer built from amino acid monomers, folded into a precise three-dimensional shape. Proteins are the workhorses of biology: enzymes, antibodies, structural materials (hair, silk), receptors, and molecular motors are all proteins. The sequence of amino acids determines the folded shape, and the shape determines the function. See also: Enzyme, Monomer, Polymer, Receptor.

Proton. A positively charged particle in the nucleus of an atom. The number of protons in a nucleus defines which element the atom is. In chemistry, "proton" also has a second, everyday meaning: a hydrogen ion (H+) in solution, a lone proton with no electron. The movement of protons is central to acid-base chemistry and to the energy-generating machinery inside cells. See also: Atom, Electron, Ion.

PVC (polyvinyl chloride). One of the world's most widely produced plastics is made from vinyl chloride monomers. The chlorine atoms in PVC's structure make it naturally flame-resistant but also problematic to recycle and incinerate, since burning it can release toxic chlorine compounds. PVC is found in pipes, flooring, window frames, and many other building materials. See also: Monomer, Polymer.

R

Racemic mixture. A 50-50 mixture of both enantiomers of a chiral molecule. Racemic mixtures form naturally in many chemical reactions because, without a chiral environment to favor one hand over the other, both mirror images are produced in equal amounts. This matters in drug development because often only one enantiomer is therapeutically useful, while the other may be inert or even harmful. See also: Chiral center, Chirality, Enantiomer, Thalidomide.

Reactant. A starting material in a chemical reaction: the substance that enters the reaction and is transformed. Reactants are consumed in the process of forming products. In photosynthesis, carbon dioxide and water are reactants. In combustion, fuel and oxygen are reactants. See also: Chemical reaction, Catalyst.

Reaction mechanism. A step-by-step description of exactly how a chemical reaction proceeds at the molecular level: which bonds break first, which atoms move, and in what order. Understanding reaction mechanisms allows chemists to predict products, design new reactions, and understand why drugs work or fail. In this book, mechanisms are told as stories rather than symbolic arrow sequences.

Receptor. A protein, usually on the surface of a cell or inside it, that binds to specific molecules and triggers a response. Receptors are the body's communication system: the molecule that binds is the message, and the receptor is the receiver that reads it and acts. Drugs work almost entirely by interacting with receptors. See also: Agonist, Antagonist, Drug, Protein.

S

Saturated vs. unsaturated. Terms describing the carbon-carbon bonds in a molecule. A saturated fat or molecule has only single bonds between carbons, meaning those carbons are "saturated" with hydrogen atoms. An unsaturated molecule has at least one double bond, which removes two hydrogen atoms and introduces a kink in the chain. This structural difference explains why saturated fats are solid at room temperature and unsaturated oils are liquid. See also: Bond (single, double, triple), Fatty acid.

Skeletal formula. The compact, line-angle drawing that chemists use to represent organic molecules. In a skeletal formula, each corner or end of a line represents a carbon atom, hydrogens bonded to carbons are omitted, and heteroatoms are written out. Once you learn to read it, a skeletal formula conveys an enormous amount of information in a very small space, like sheet music for molecules. See also: Carbon, Heteroatom, Molecule.

Solvent. A liquid that dissolves another substance (the solute) to form a solution. Water is the universal solvent for biology. Many organic reactions require organic solvents because the molecules involved don't dissolve in water. One of green chemistry's priorities is finding safer, more sustainable solvents. The choice of solvent can dramatically affect how a reaction proceeds. See also: Green chemistry.

Space-filling model. A three-dimensional representation of a molecule in which each atom is shown as a sphere sized according to its actual atomic radius, and spheres overlap where atoms are bonded. Space-filling models show what a molecule would look like if you could see its electron clouds. They are less useful for showing connectivity than skeletal formulas, but much better for understanding how a molecule might fit into a receptor pocket.

Spectroscopy. A family of techniques that use light (or other forms of radiation) to identify and characterize molecules. Different types of spectroscopy reveal different things: infrared spectroscopy shows which functional groups are present; mass spectrometry measures molecular mass; NMR spectroscopy maps the arrangement of atoms. Spectroscopy is how chemists determine the structure of a new molecule without physically disassembling it.

Stereochemistry. The branch of chemistry concerned with the three-dimensional arrangement of atoms in molecules and how that arrangement affects properties and reactivity. Chirality is the most famous topic in stereochemistry, but the field also covers the geometry around double bonds and the shape of rings. Two molecules with identical connectivity but different three-dimensional shapes can have completely different biological effects. See also: Chirality, Enantiomer.

Substitution reaction. A reaction in which another replaces one atom or group in a molecule. Think of it as a molecular swap: one visitor leaves, and a new one takes its seat. Substitution reactions are workhorses of organic synthesis and of biological chemistry, used constantly to modify molecules and introduce new functional groups. See also: Chemical reaction, Elimination reaction, Reaction mechanism.

Sugar. A broad term for a family of sweet, water-soluble carbohydrates. Simple sugars (monosaccharides) like glucose and fructose are single-ring-shaped molecules. Complex sugars (polysaccharides) like starch and cellulose are polymers of simple sugars. Sugars are the primary fuel source for most living things, and they play important structural roles in DNA and in plant cell walls. See also: Cellulose, Glucose, Polymer.

Synthesis. In chemistry, the process of building a target molecule from simpler starting materials through a planned sequence of reactions. Total synthesis means building a complex natural product entirely from scratch, using simple, inexpensive precursors. Drug synthesis is the industrial process by which pharmaceutical companies reliably manufacture a drug molecule at scale. See also: Chemical reaction, Catalyst, Natural product.

Synthetic. Made in a laboratory or chemical plant rather than extracted from a natural source. Synthetic does not mean inferior: a molecule synthesized in a lab is chemically identical to the same molecule from a natural source. The vanilla flavor in most commercial products is synthetic vanillin, identical to the vanillin in a natural bean. See also: Natural product, Synthesis, Vanillin.

T

Terpene. A large family of organic molecules built from repeating five-carbon units, produced in abundance by plants and some insects. Terpenes include many of the most recognizable smells in nature: limonene (citrus), pinene (pine forests), and linalool (lavender). Cholesterol and many vitamins are technically terpenes too, illustrating how far this chemical family extends. See also: Functional group, Natural product.

Tetrahedral geometry. The three-dimensional shape adopted by a carbon atom that is making four single bonds. The four bonded atoms point toward the corners of a tetrahedron, with bond angles of about 109.5 degrees. This geometry is central to chirality: a carbon at the center of a tetrahedron with four different groups attached is a chiral center. See also: Bond (covalent), Chiral center, Chirality.

Thalidomide. A drug prescribed in the late 1950s primarily in Europe as a sedative and anti-nausea medication for pregnant women was later found to cause severe congenital disabilities in thousands of children. The tragedy traced directly to the drug's chirality: one enantiomer of thalidomide had the intended therapeutic effect, while the other caused developmental harm. Thalidomide's story permanently changed how the pharmaceutical industry thinks about chiral drugs and safety testing. See also: Chirality, Enantiomer, Chiral center, Drug.

Toxicity. The degree to which a molecule harms a living organism. Toxicity depends on dose, route of exposure, and the specific biological target the molecule interacts with. Many molecules that are beneficial at low concentrations become toxic at high concentrations. Organic chemistry offers both tools for understanding toxicity at the molecular level and methods for designing safer molecules. See also: Drug, Receptor.

Triglyceride. A molecule consisting of three fatty acid chains attached to a glycerol backbone. Fats and oils in food are primarily triglycerides. The length and saturation of the fatty acid chains determine whether a triglyceride is solid (like butter, with saturated chains) or liquid (like olive oil, with unsaturated chains) at room temperature. See also: Fatty acid, Saturated vs. unsaturated.

V

Vanillin. The primary flavor molecule in vanilla is responsible for its characteristic warm, sweet aroma. Vanillin is an aldehyde with a phenol group and a methoxy group on an aromatic ring. The vanillin molecule from a natural vanilla bean is chemically identical to vanillin synthesized in a laboratory. This equivalence sits at the heart of the "natural vs. artificial" debate in food labeling. See also: Aldehyde, Aromatic ring, Phenol, Synthetic.

Y

Yeast. Single-celled fungi are among humanity's oldest chemical collaborators. Yeast drives fermentation, converting sugars into ethanol and carbon dioxide through a series of enzyme-catalyzed reactions. They are used in brewing, baking, and increasingly in biotechnology to produce chemicals, fuels, and pharmaceuticals. At the molecular level, yeast is a miniature chemical factory, running complex organic transformations with remarkable efficiency. See also: Enzyme, Ethanol, Fermentation.

13 References

Atkins, Peter, and Julio de Paula. *Elements of Physical Chemistry*. 9th ed., Oxford University Press, 2023. (CH1)

Berg, Jeremy M., et al. *Biochemistry*. 10th ed., W. H. Freeman, 2024. (CH1)

Bergin, Edwin A. *Life in the Universe: Expectations and Constraints*. MIT Press, 2023. (CH1)

Borman, Stu. "Graphene and Beyond: Carbon Allotropes in Modern Materials Science." *Chemical & Engineering News*, vol. 102, no. 14, 2024. (CH1)

Breslow, Ronald. "The Origin of Homochirality Revisited." *Accounts of Chemical Research*, vol. 57, no. 3, 2024. (CH1)

Brown, Theodore L., et al. *Chemistry: The Central Science*. 16th ed., Pearson, 2023. (CH1)

Cao, Yiqing, et al. "Advances in Carbon-Based Nanomaterials for Energy Storage." *Nature Materials*, vol. 23, 2024. (CH1)

Crespo-Hernández, Carlos E., et al. "Ultrafast Photochemistry of Organic Molecules." *Chemical Reviews*, vol. 124, no. 9, 2024. (CH1)

Doudna, Jennifer A., and Samuel H. Sternberg. *The Future of Genetic Engineering*. HarperCollins, 2024. (CH10)

Dunn, Patrick J. "Green Chemistry: Foundations and Frontiers." *Green Chemistry*, vol. 26, no. 1, 2024. (CH10)

Emsley, John. *Nature's Building Blocks: An A–Z Guide to the Elements*. Updated ed., Oxford University Press, 2023. (CH1)

Fleming, Ian. *Molecular Orbitals and Organic Chemical Reactions*. 2nd ed., Wiley, 2023. (CH1)

Gates, Bruce C., et al. "Catalysis and Carbon Transformations." *Science*, vol. 383, 2024. (CH9)

Grammarly AI Editor, 2026 (CH1 – CH12)

Ghosh, Abhik, and Jeanet Conradie. "Carbon Bonding and the Limits of Organic Stability." *Journal of Organic Chemistry*, vol. 89, no. 2, 2024. (CH1)

Hawkins, Thomas R., et al. "Life-Cycle Carbon Analysis of Modern Materials." *Environmental Science & Technology*, vol. 58, no. 11, 2024. (CH9)

Hoffmann, Roald. *The Same and Not the Same*. Updated ed., Columbia University Press, 2023. (CH1)

Holliday, Gemma L., et al. "Carbon Metabolism Pathways in Modern Biology." *Nature Reviews Chemistry*, vol. 8, 2024. (CH1)

Kanan, Matthew W. "Carbon Fixation Chemistry: Artificial Pathways." *Nature Chemistry*, vol. 16, 2024. (CH9)

Krebs, Robert E. *The History and Use of Our Earth's Chemical Elements*. 4th ed., Greenwood, 2023. (CH1)

Lerner, Louise. "New Insights into Carbon's Stellar Formation." *Physical Review Letters*, vol. 132, 2024. (CH1)

Li, Xiaoqing, et al. "Graphene and 2D Carbon Materials in Electronics." *Advanced Materials*, vol. 36, 2024. (CH1)

Liu, Yifan, et al. "Organic Reaction Mechanisms Without Catalysts." *Chemical Science*, vol. 15, 2024. (CH5)

Miller, Stanley, and Jeffrey L. Bada. *The Origin of Life Revisited*. University of Chicago Press, 2023. (CH1)

Nocera, Daniel G. "Artificial Photosynthesis and Carbon Recycling." *Accounts of Chemical Research*, vol. 57, no. 12, 2024. (CH9)

Olah, George A., et al. *Beyond Oil and Gas: The Methanol Economy*. 4th ed., Wiley-VCH, 2023. (CH9)

Pilling, Michael J., and Paul W. Seakins. *Reaction Kinetics*. 3rd ed., Oxford University Press, 2023. (CH5)

Ritter, Stephen K. "Carbon's Versatility in Modern Organic Synthesis." *Chemical & Engineering News*, vol. 102, no. 22, 2024. (CH1)

Schmidt-Rohr, Klaus. "Why Carbon Chains Dominate Organic Structures." *Journal of Chemical Education*, vol. 101, no. 4, 2024. (CH1)

Seager, Sara. *The Smallest Lights in the Universe: Carbon and Cosmic Chemistry*. Updated ed., Crown, 2024. (CH1)

Sherratt, Thomas N. *The Evolution of Life's Chemistry*. Princeton University Press, 2023. (CH1)

Smith, Janice G. *Organic Chemistry*. 7th ed., McGraw-Hill, 2024. (CH1)

Solomon, Edward I., et al. "Carbon in Biological Catalysis." *Nature Reviews Chemistry*, vol. 7, 2023. (CH1)

Turchi, Craig S., et al. "Carbon Capture Technologies: A 2024 Review." *Energy & Environmental Science*, vol. 17, 2024. (CH9)

Zhang, Wei, et al. "Chirality and Molecular Architecture." *Nature Chemistry*

www.ingramcontent.com/pod-product-compliance
Lightning Source LLC
LaVergne TN
LVHW022154200926
848481LV00006B/1740